SPACE TODAY CHRONICLES

VOLUME ONE

THE ONGOING STORY OF WHAT'S HAPPENING IN SPACE TODAY
FROM THE EDITORS OF *SPACE TODAY* MAGAZINE

Zachary Luhrs

SPACE TODAY CHRONICLES

VOLUME ONE

THE ONGOING STORY OF WHAT'S HAPPENING IN SPACE TODAY
FROM THE EDITORS OF *SPACE TODAY* MAGAZINE

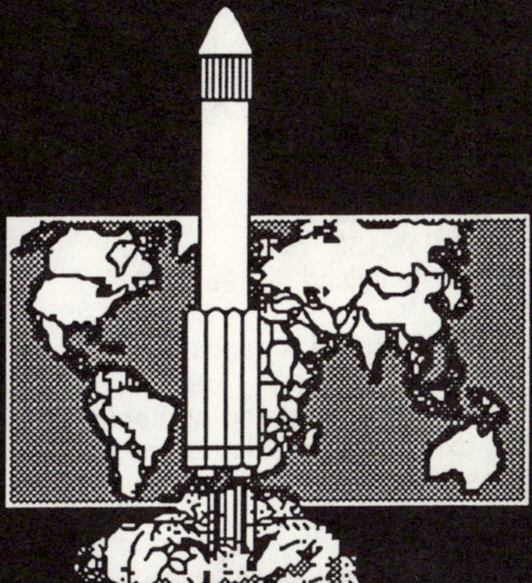

WRITTEN AND EDITED BY
ANTHONY R. CURTIS
EDITOR OF *SPACE TODAY* MAGAZINE

ARCSOFT PUBLISHERS
WOODSBORO, MARYLAND

VOLUME ONE
1987 EDITION
FIRST PRINTING

This 1987 Edition book is Volume One in an annual series covering Space from Earth to the edge of the Universe. The next volume will be Volume Two, the 1988 Edition. Readers requiring mid-year updates should consult *SPACE TODAY* monthly newsmagazine published by ARCsoft Publishers.

Postoffice Box 132
Woodsboro Maryland 21798 USA

Printed in the United States of America.

Satellite data in this book is based on information from the National Aeronautics and Space Administration, Goddard Space Flight Center and other NASA centers, NORAD, the Smithsonian Astrophysical Observatory, European Space Agency, satellite owners and many other sources.

ISBN 0-86668-064-0

Preface

If you're like me, you scan the local newspaper everyday, hoping for a tidbit about the latest deep space discovery by astronomers, radio signal from Uranus and Neptune, Russian space shuttle or U.S. shuttle flights, scientific satellite, space station in orbit, or unexpected supernova. You check out tv news for scraps of information on men in space from Russia or the latest Chinese, Japanese and European launches. But you find precious little to satisfy that deep hunger to know what's going on Out There.

An x-ray glow has been seen around the core of our Milky Way galaxy. Building bricks can be made from Moon dust. The edge of the Universe we know has been extended by finding a distant quasar. Some say the Russians are 15 years ahead of the U.S. in manned flight experience, having already occupied a second-generation space station in orbit. Do the networks report any of this news? Frustratingly, no!

That's why this book exists. There have been, continue to be, and will be so many developments affecting mankind's future, they simply are bursting to be out of the news box. This book rounds up the year's major news about space stations, shuttles, unmanned satellites, planetary explorers, the Moon, our Sun, Mars, Venus, Saturn, Jupiter, our entire Solar System, our Milky Way galaxy and neighboring galaxies, Deep Space, pulsars, quasars, the Universe, countries like Russia, Japan, China and Europe in Space, cosmology, technology. It includes updates and background checks on items too small for the popular press to report.

From the Sun to Pluto, from the Milky Way to pulsars and quadruple stars, this book will bring you up to date with what's happening just now Out There. *—ARC*

Table of Contents

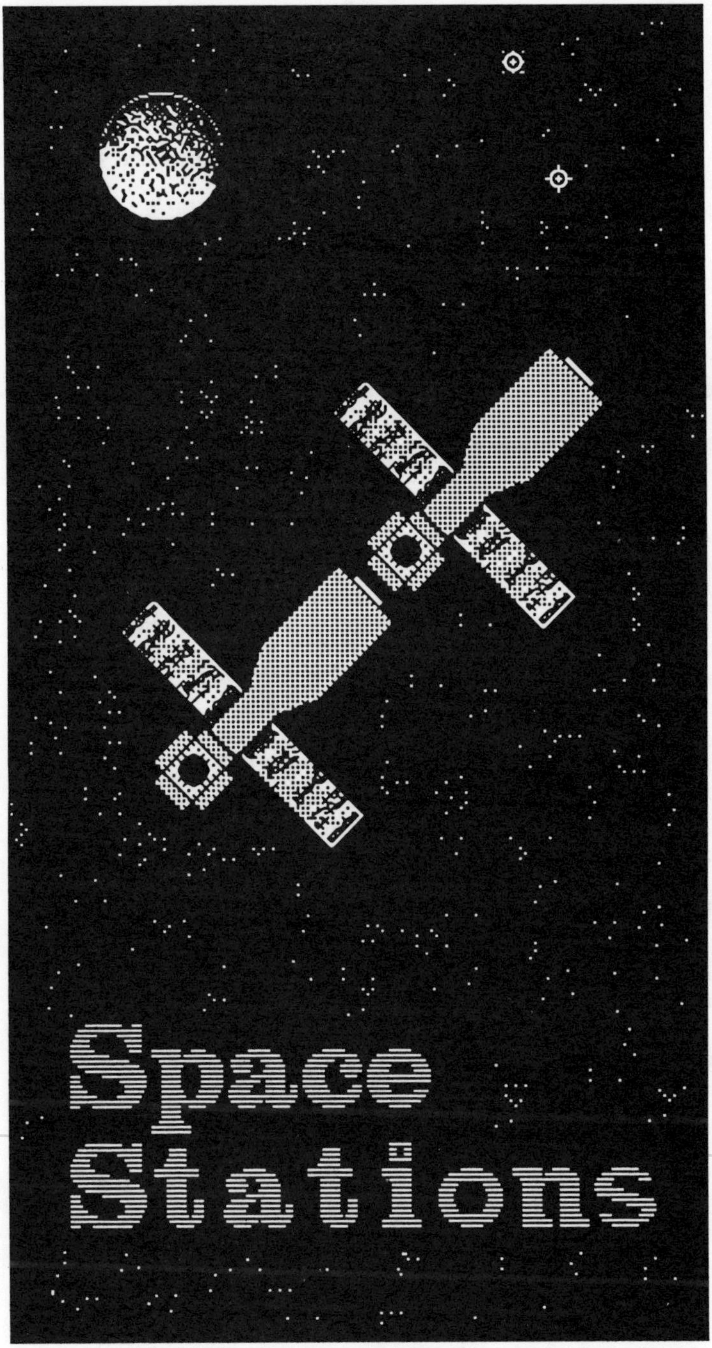
Space
Stations

Russia's *Mir* Space Station

Against a backdrop of grounded U.S. space shuttles and European *Ariane* rockets, the USSR is making space history:

✔ launching 90 to 100 satellites a year into Earth orbit,

✔ adding building-block modules to the *Mir* space station in orbit,

✔ placing men permanently in space,

✔ and flying satellites to orbit for other countries at bargain-basement prices.

History may recall 1987 as the year men went to space permanently. *Mir,* the second-generation space station, was placed in orbit by the USSR in 1986. Men worked in the station, on and off during 1986, getting it ready for permanent occupancy. Cosmonauts lived aboard *Mir* to test the new hardware until Fall 1986, then came back to Earth to make final preparations for a permanent presence in space aboard.

Progress 27, an unmanned Earth-to-space freighter loaded with food, fuel and supplies for the space station, was launched in subzero temperatures January 16, 1987.

It parked in an orbit ranging from 114 miles out to 164 miles. Two days later it was flown out to 200 miles for a rendezvous and docking with *Mir*. The docking was broadcast live to USSR television audiences. Little was seen on U.S. tv.

Progress 27 pushed the combined freighter and space station to a circular orbit 215 miles above Earth which permits launch windows from the Soviet Union every two days.

On February 6, 1987, two cosmonauts, Yuri Romanenko and Alexander Laveikin, were launched in a new-style *Soyuz TM-2* capsule to board *Mir*. After connecting *Soyuz* into a port on Mir, the cosmonauts went about unloading provisions from *Progress 27* into *Mir*.

Permanent crew. When those cosmonauts entered *Mir* in February 1987, it seemed likely there would be an unbroken string of men in space forevermore.

In April 1987, space laboratory module *Kvant* was sent up and attached successfully to *Mir,* increasing the working room for cosmonauts.

Mir is 43 ft. long, big as a house trailer in orbit. Modules of similar size are expected to be ferried up from time to time, docked with *Mir,* and permanently installed as part of the space station.

Russia's first space station design was *Salyut*. Several versions of that spacecraft have been used successfully in space since the early 1970s.

Mir is more like a home than the first-generation *Salyut* space station was. The galley, for instance, has been made more palatable.

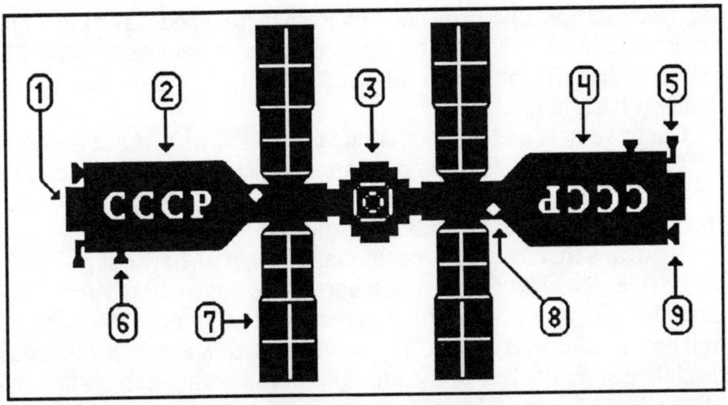

USSR *MIR* SPACE STATION ② grows with addition of a second *Mir* module ④ in this artist's conception. Large six-port dock ③ attached to the small end of *Mir* accepts up to five spacecraft. Single-port dock ① is on the large end of *Mir.* A large antenna ⑤ sends signals to a communications relay satellite which retransmits signals to ground controllers. Small antenna ⑥ guides approaching spacecraft to dock. Solar panels ⑦ convert sunlight to electricity to power the space station. The small main engine ⑨ moves the station if needed. Cosmonauts inside have individual sleeping compartments, work stations, a bicycle exerciser, a treadmill exerciser, large toilet and washing facility, and a window ⑧ with blind.

Special private compartments for crew members have been included along with nicer shower and bathroom facilities.

The 42,000-lb. *Mir* has docking ports on each end. *Soyuz* is attached to a single port at the large end of *Mir*. One of six ports on a docking hub is attached to the small end of *Mir*.

That docking hub is a sphere with six ports where spacecraft are hooked up. One port attaches the docking hub to the small end of *Mir*. The other five ports are available for attaching other spacecraft to the original *Mir* craft.

Restocking the larder. USSR flight controllers have been sending *Progress* unmanned Earth-to-space freighters loaded with food, fuel and supplies to the station about once a month.

Docking of *Progress* 27, launched in subzero temperatures in January 1987 from Baikonur Cosmodrome, was broadcast live to USSR television audiences.

Several other *Progress* freighters have ferried more supplies to the space station since cosmonauts Romanenko and Laveikin went up. *Progress 30,* for instance, docked at the space station May 21, 1987, with a fresh supply of food, water, fuel and mail from relatives back home.

Cargo vessels and manned flying craft dock with *Mir* at ports on either end of the 43-ft. station. A single dock is at the large end of the station. A multiple-port dock capable of receiving five other spacecraft is at the small end of the station.

Space docks. The cosmonauts keep their *Soyuz TM-2* space flyer attached to the single port at the large end of *Mir*. They park their *Progress* unmanned cargo ships at ports in the docking hub on the other end of *Mir*. The newly-arrived *Kvant* space-science module also is docked at the multiple-port docking hub on the end of *Mir*.

Kvant and other add-on modules fly unmanned up to the *Mir* space station. When a module arrives at the space station, it docks temporarily at the central port opposite *Mir* on the hub. After checkout by the station crew, a manipulator arm grabs the module and swings it around to a port on the side of the docking hub. In that position, it sticks out like a spoke at a right angle from *Mir*.

Various 42,000-lb. 43-ft. modules, duplicates of *Mir,* will be flown to orbit from time to time as add-ons loaded with equipment to increase the size of the space station.

Ultimately, six *Mir* craft can be docked together to create a manned 250,000-lb. six-pointed "star" in space measuring 85-90 ft. in diameter. Another such six-pack could be mated to the single port on the large end of each *Mir* creating an ever-growing complex city of *Mir* modules in space.

From now on. Cosmonauts Romanenko and Laveikin started man's permanent residence in space. They are to be visited by other cosmonauts in the summer of 1987 and relieved for a trip back down to Earth after about 290-300 days in orbit, breaking the old 237-day record for manned space flight. A crew including at least one USSR cosmonaut and one from Syria is expected to go to *Mir* in July 1987.

Crews do scientific research, commercial development work and military projects. They are to be changed every ten months or so.

For the first time, men have enlarged permanently their home in space. The *Kvant* astrophysics module is 19 ft. long and 13.6 ft. in diameter. It is permanently attached to one of *Mir's* docks, greatly increasing the working room for cosmonauts.

Kvant, is an astronomer's delight—an x-ray observatory including electronic equipment from the European Space Agency.

The Soviet Academy of Sciences, the European Space Agency, and West Germany each supplied special spectrometers. Great

Britain and the Netherlands supplied a telescope for electronic images.

An observatory above Earth's clouds and dusty atmosphere sees the Moon, Sun, planets, Milky Way galaxy and deep space much more clearly.

An Earth-resources module, planned for launch in the summer of 1987 and similar in size to *Mir*, will increase again the residential roominess of the station. Photographing mineral and fuel resources on Earth's surface from space could offer large economic benefits to the Russians.

Earth-resources photos made in 1981 and 1982 by cosmonauts in

Salyut 6 and *Salyut 7* space stations reportedly brought several million dollars in benefits to the USSR economy.

The Russians have built a chain of five ground stations to receive valuable data from Earth-resources satellites. A new antenna in Alma Ata joins stations in Moscow, Tashkent, Novosibirsk and Khabarovsk receiving signals from *Meteor* weather satellites and other craft transmitting pictures of Earth from space.

Shuttle to space city. The USSR may be able to loft cosmonauts to its growing space complex in a new space shuttle expected to be test-fired by early 1988.

And a huge new heavy-lifter rocket capable of carrying more than 110 tons to orbit was test fired successfully in May 1987 from Baikonur Cosmodrome. The booster delivers 6.6 million-lbs. of thrust, similar to *Saturn V* which lifted Americans to the Moon in 1969-72. A manned flight with the new Russian shuttle attached could come late in 1987 or early in 1988.

Superficially, the Russian space shuttle looks like the U.S. shuttle. But, unlike the U.S. shuttle, the USSR shuttle has no built-in rocket motors. Rather, the main engines are in the bottom of the huge rocket core. Four liquid-fuel booster rockets are strapped around the core. A shuttle or an unmanned cargo container can be strapped to the core for flight to orbit.

The powerful new 100-ton lifter also could be used to carry giant freight loads to low-Earth orbits 500 miles high, heavy-weight satellites to stationary orbits at 22,000 miles out, or men and materials to the Moon or Mars.

The USSR's space shuttle already has been mounted for ground testing on the side of the huge rocket on its new launch pad beside a 3.5-mile shuttle landing runway, roads, buildings and railways at the Baikonur Cosmodrome in the eastern part of the the USSR state of Kazakhstan.

The USSR also may be preparing a smaller space plane to fly only people to orbit and back, as well as a larger space truck like the U.S. shuttle to carry both men and cargo.

By turning the lights on permanently in the space station, turning up the heat in the international commercial launch business, moving ahead on space research and manufacturing, lofting scores of satellites to orbit, and flying more powerful rockets and shuttles, the USSR can only increase its influence in space in the future.

Spacewalk Saves Science Lab

Russian cosmonauts took a dramatic walk in '"raw space" from the *Mir* orbiting station April 11, 1987, to remove a small white

cloth bag which blocked a major science module from docking with their laboratory.

Soviet television, broadcasting from mission control outside Moscow, showed the first live broadcast of a Soviet space walk as Yuri Romanenko and Alexander Laveikin left *Mir* and floated to the docking port attached to the astrophysics module *Kvant*.

Launched March 31, 1987, *Kvant* is a biotechnology and astrophysics space laboratory. It has an observatory, life-support systems, radio, telephone, telegraph and television equipment.

April 5, *Kvant* was moved within 200 yards of the *Mir* space station but could not dock due to a mysterious control-system problem.

April 9, another docking attempt was made. It touched Mir but the two spacecraft were not firmly linked. The joining was not airtight. The *Mir* crew could not walk through a tunnel in the docking port to the *Kvant* module and "get down to astrophysical research," as Radio Moscow put it.

The spacewalk. April 11, Romanenko and Laveikin went into what Radio Moscow called "raw space" to find the problem and fix it. They were successful. The airtight linkage permits cosmonauts to move freely between *Mir* and the 11-ton *Kvant.*

Romanenko and Laveikin spent 3 hours 40 minutes outside *Mir,* floating in space, removing what they called an "alien object."

Laveikin spotted the small white cloth bag stuck between the craft. He radioed mission control to move *Kvant* back a bit. Ground controllers ordered small thrusters in *Kvant* to fire, backing it off a few inches from the dock so Laveikin could grab the bag. Romanenko reported the bag was removed and the joint between the two craft was clear for a solid docking. He asked mission control to pull *Kvant* up to the dock.

April 13, Romanenko and Laveikin opened the hatch to inspect *Kvant* and remove grease safety coatings from its research equipment so they can conduct experiments in the module.

USSR newspapers carried detailed accounts of the spacewalk and Radio Moscow broadcast the conversation between cosmonauts and ground control. Romanenko and Laveikin went to space February 6.

The *Kvant* astrophysics module is 19 ft. long and 13.6 ft. in diameter. It will be kept attached permanently to one of *Mir's* docks, increasing the working room for cosmonauts.

The new space-science research addition is an astronomer's delight—an x-ray observatory including electronic equipment from Western Europe's space agency. The Soviet Academy of Sciences, the European Space Agency, and West Germany each supplied special spectrometers. Great Britain and the Netherlands supplied a telescope for electronic images. An observatory above Earth's clouds and dusty atmosphere sees the Moon, Sun, planets, Milky Way galaxy and deep space much more clearly.

Space Station: Lab or Factory?

When the U.S. second-generation space station goes into use in ten years, will it house only science research labs and a military base? Or will it include a large private manufacturing area where drugs and metal alloys are made for profit?

With the station on the drawing board in the aftermath of the shuttle *Challenger* disaster, government leaders, businessmen, scientists and the military are hashing over what the space-home-away-from-home will be used for.

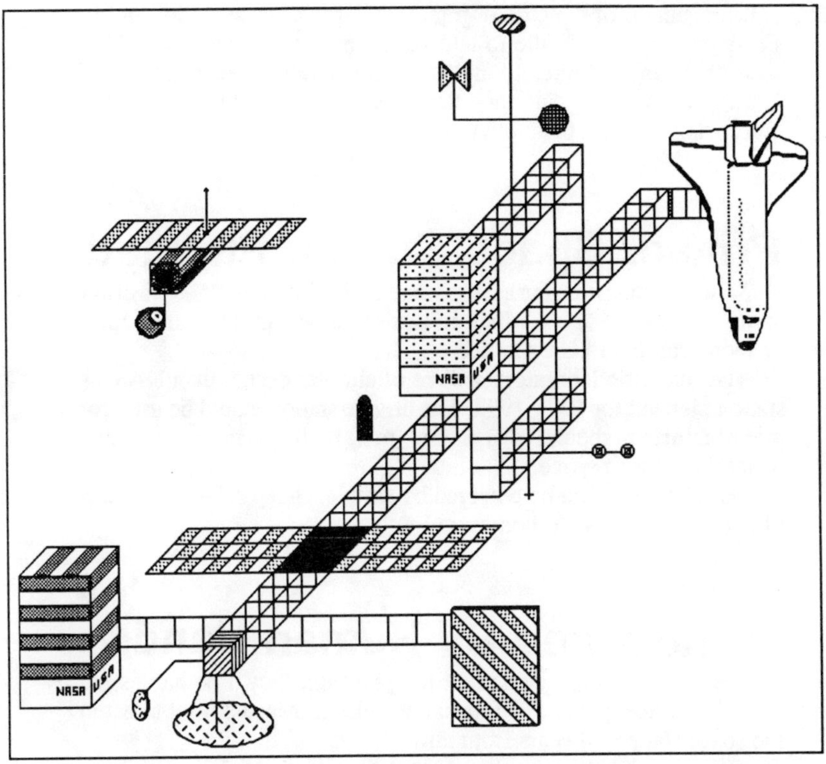

Some political leaders argue the space station should be reserved for pure scientific and military uses. Commercial pressures from private businesses already have forced too intense a launch schedule for shuttles. Commercial uses of space should be carried out in privately built and launched satellites and manned space stations, they say.

Some businessmen counter that they want to use the government's station as a factory in orbit. One recent prediction said space manufacturing could be a $100 billion a year business by the year 2005. Business profits would increase if companies did not have to pay for a separate station in orbit.

Even more fantastic electronics than we now have will be needed. Every machine in a space factory would be automated. Factories packaged in 2,500 cubic foot modules attached to the main station would be run entirely by microcomputers.

The USSR already has its *Mir* space station in orbit. U.S. plans call for launch of construction materials for a space station in low Earth orbit in the middle to late 1990s. Space shuttles would ferry astronauts and supplies to and from the station. The Japanese and Europeans are designing modules to be attached to the U.S. station. Europeans also are planning a station of their own.

Private Space Station Planned

Space Industries Inc., Houston, is building a $500 million unmanned space station which it hopes will be carried aloft in 1989 or soon after by a U.S. shuttle.

The scientific laboratory will be of simpler design than NASA's station planned for 1995-1999. The private station would be used for manufacturing special drugs and metal alloys in the relative weightlessness of space.

The station would be powered by simple, cheap solar-cell panels like those used in solar homes on Earth.

Untouched By Human Hands

Tracy and Harvey won't be tilling soil but they'll be harvesting crops of lettuce, potatoes and whatever else astronauts want to eat in the space station 20 years from now.

Tracy will be a tray-handling robot and Harvey will be a harvesting robot in a farm module attached to the U.S. space station after the turn of the century.

Human viruses and funguses carried by astronauts could wipe out crops, leaving spacemen with nothing to eat. Sanitary robots, on the other hand, could do the time-consuming chores, protect the crops, and let astronauts get on with other business.

Believe it or not, it will cost about three-quarters of a billion dollars to send the farm to space. But that's still cheaper than continuously carrying brown-bag lunches up to workers in space.

The compact farm would be about the size of a school bus, powered by an array of solar cells converting sunlight to electricity, and use fiber optics to spray sunlight directly on plants flowering in Tracy's trays of water.

Harvey and his sidekick will plant, tend and harvest, calling for help only if something breaks down. Humans won't be allowed near the crops except in emergencies.

Each robot, equipped with trowel and other gardening tools, would roll up and down the length of the farm module on a 45-ft. track, tending its crops.

Superman's Hand

An aluminum hand with five natural-looking fingers and the strength of a superman is being tested by NASA for use in the 1990's space station.

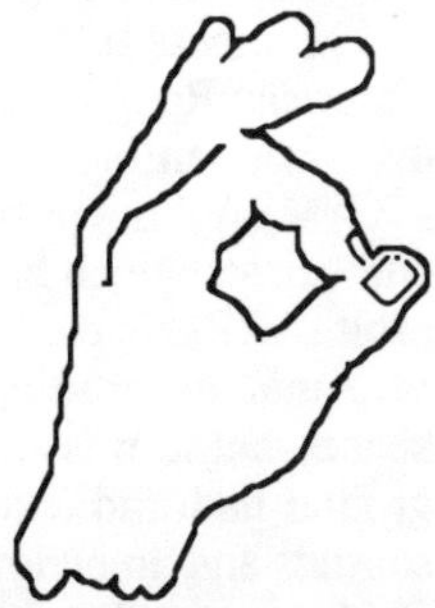

The hand has motorized metal links which curl the fingers around an object and hold on with great strength. In space it will help build the U.S. station planned for the late 1990s, grab broken satellites from orbit and maybe even save astronauts floating away from home base. And, if it works in space, it undoubtedly will lead to workable hands for handicapped persons on Earth.

Space station construction has been budgeted at $8 billion. One-tenth of the money must be spent on artificial intelligence and robotics. The artificial hand is part of that research.

The station isn't expected to be in orbit until 1993 at the earliest. A dozen shuttle flights will ferry men and materials to orbit to build the station. When ready, a crew of six to eight astronauts will live there continuously in rotating shifts. The station will provide a low-gravity science lab, a base for military intelligence gathering and, maybe, a commercial factory.

Shuttle Space-Repair Bay

Service industries are growing in the United States and astronauts probably will take them to space as satellite maintenance and repair becomes a major job for U.S. shuttle astronauts.

U.S. shuttles will not be placing many private commercial satellites in orbit after they go back to work in 1988. Government policy calls for mostly scientific research and military work. One special line of work will be repair of military and science satellites in orbit.

To facilitate repairs of all types of satellites in space, Lockheed and NASA are working up designs for new repair rigs called *SAMS* for Space Assembly Maintenance and Service.

In 1995 they plan to fly to space, in the shuttle cargo bay, a new rig which will capture and hold satellites during repairs. By the year 2010 a robot service station will be carried to orbit by a shuttle. The robot will fly itself around to repair objects in orbit.

Ham Radio On Space Station

There will be a permanent ham shack aboard the U.S. Space Station ten years from now if American amateur radio operators are successful in selling the idea to NASA.

Members of the Radio Amateur Satellite Corp. (AMSAT), the American Radio Relay League (ARRL) and NASA in October 1986 discussed putting a ham station aboard the space station.

AMSAT is the worldwide organization of amateur radio space enthusiasts which has put a dozen of its own ham radio satellites in orbit. ARRL is the U.S. national amateur radio club.

Amateur radio operators have been among astronauts flying on space shuttle missions in recent years. Many successful space-to-ground ham radio contacts have been completed between orbiting shuttles and amateurs in their home stations on Earth.

Astronaut Tony England, who operates his own amateur radio station under the call letters WØORE, took part in the discussions. Also supporting construction of amateur radio facilities in the coming space station was NBC television newsman Roy Neal whose amateur callsign is K6DUE.

Astronaut Tony England said amateur radio operations from space build interest among school students for science, mathematics and engineering. A ham shack aboard the space station also would provide recreation for the high-tech-oriented astronauts living there.

Stations: Europe and Japan

Columbus is a large manned scientific module to be attached to the U.S. international space station in the late 1990's and, later, to become part of an independent European space station.

Columbus can be used as an unmanned platform, a stand-alone manned module, or as a pressurized module attached to a space station.

The manned delta-wing European spaceplane *Hermes* shuttlecraft will carry two pilots and two to four other crew members to service and maintain unmanned platforms in space, manned scientific modules in space, and the planned *Columbus* European space station. It will be able to replace crew members on any space station and restock the station with food and other consumable supplies.

Europeans expect the use of large unmanned space platforms to increase. *Hermes* can carry maintenance men to those platforms to restock fuel and bring materials produced on the platforms back to Earth.

Meanwhile, Japan also is building a module to be attached to the U.S. international space station in the late 1990's. Japan will have a small space shuttle to fly up to the station in orbit.

Space
Shuttles

Russian Shuttle Testing

New U.S. spy satellite photos reportedly have revealed the USSR's space shuttle mounted for testing on the side of a huge rocket on a new launch pad at Tyuratam Spaceflight Center.

The orbiter reportedly was hauled away from the pad after tests. The USSR may flight test the one-time-use rocket with only a cargo pod in 1987. The shuttle may be ferried to orbit for the first time early in 1988.

A shuttle landing strip has been under construction at the Baikonur/Tyuratam Spaceflight Center in the eastern part of the USSR state of Kazakhstan. Shuttle landing runways, roads, buildings and railways were seen in 1986 in photos from Sweden.

The runway may be 25 percent longer than the runway at Kennedy Space Center in Florida. And the building where a shuttle is attached to a rocket may be larger than one in Florida. The photos released in Sweden, from a satellite 500 miles above the Baikonur Space Center near Leninsk in Kazakh SSR, show a 3.5 mile runway and a giant vehicle assembly building. The Kennedy shuttle runway, by comparison, is only 2.8 miles long. The USSR vehicle assembly building appears to be half again as large as a similar facility at Kennedy.

It's not clear whether the USSR is preparing a small space plane to fly only people to orbit and back, or a larger space truck to carry both men and cargo like the U.S. shuttle.

Japanese Shuttle

Everybody wants a truck to haul stuff out into space and bring it back. The Russians are rushing to build a shuttle. The Europeans are planning one. Now the Japanese are studying how to build a shuttle of their own.

Japan's Science and Technology Agency is looking into designs, construction techniques and the question of whether to go it alone or throw in with another country.

They could build a small space plane to be launched atop the powerful Japanese H-2 rocket expected to be available for use in 1991. A larger shuttle could be built later and launched on an H-2 with extra booster rockets strapped to its sides.

Eventually, the Japanese want a combination airplane and space plane which will take off from, and land on, runways on Earth. That development is expected after the turn of the century.

European Shuttle

The European spaceplane *Hermes*, a manned delta-wing shuttlecraft, is closer to reality now that 13 European countries and Canada have put up money to get started.

Design work has been started on propulsion, electronics, fuel cells, guidance control, a manipulator arm, temperature and environment control, life support, computers, communications, and an airlock for spacewalks.

Hermes will be about 59 feet long with a wingspan near 33 feet. It will carry 10,000 pounds of payload in a 1,236 cubic foot cargo bay about ten feet across. Plans call for *Hermes* to be ferried to orbit by an *Ariane 5* rocket. It will fly back down and land on a runway.

Hermes will carry two pilots and two to four other crew members to service and maintain unmanned platforms in space, manned scientific modules in space, and the planned European space station. It will be able to replace crew members on any space station and restock the station with food and other consumable supplies.

The European Space Agency decided last year to start planning the shuttle. A final decision to build the spaceplane is expected this year. Europeans expect the use of large unmanned space platforms to increase. *Hermes* can carry maintenance men to those platforms to restock fuel and bring materials produced on the platforms back to Earth.

Columbus is a large manned scientific module to be attached to the U.S. international space station in the late 1990s and, later, to become part of an independent European space station. *Columbus* can be used as an unmanned platform, a stand-alone manned module, or as a pressurized module attached to a space station.

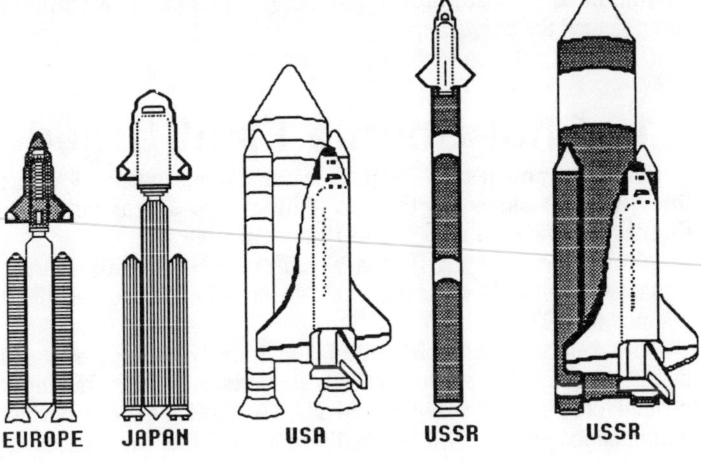

Challenger Disaster Aftermath

After the U.S. shuttle *Challenger* exploded enroute to orbit in January 1986, U.S. space plans were in disarray and satellite owners were left with ready-to-fly birds on the ground.

Shuttle *Columbia* would have been up and back in July 1986 with Western Union's *Weststar 6* radio-TV satellite. Unmanned probes would be flashing toward Jupiter and the Sun. But NASA had no way to launch satellites until shuttle flights resume late in 1988. And the Europeans had their own rocket failures to investigate.

It's became something of a national embarrassment with private U.S. satellite owners scrambling to rent space atop a Chinese *Long March* rocket. Several satellite builders talked to the Chinese about *Long March* launches. Teresat, a Houston company, contracted with China.

Meanwhile the USSR wanted to pick up hard dollars by selling excess launch capacity. The Japanese launched satellites in 1986 but they were not in a position to fill the US and European launcher shortage.

The shuttle was everywhere in U.S. plans. NASA did not want unmanned rockets which could be used only once. But after the *Challenger* explosion, many called for a fleet of unmanned expendable rockets to place satellites in Earth orbit. With no launch capacity, the U.S. appeared through 1986 and 1987 to be losing ground as the leading spacefaring nation.

NASA wanted to resume its role as the official national rocket launching agency. But private enterprise backers wanted corporations to get into the launching business. The Air Force wanted to build and launch its own unmanned rockets capable of carrying spy and communications satellites to orbit. Indecision in Washington compounded the confusion.

13-Shot Shuttle Limit Urged

The government's National Research Council wants NASA to limit itself to eleven to thirteen shuttle trips a year when four shuttles are available after flights resume in 1988.

NASA has said 16 flights a year will be possible when a replacement shuttle is in use. That's the same number originally planned for 1986.

The research council called for improvements to shuttle maintenance facilities, addition of spare parts, and better training for ground crews. The council said NASA should have a periodic maintenance program which could ground shuttles when necessary, just as airlines manage their jet fleets.

NASA is planning five shuttle flights in 1988, ten in 1989 and eleven in 1990.

Rebuilding NASA

NASA officials said at the beginning of 1987 they are getting their act together again, a year after the *Challenger* disaster. Here's their list of good things happening:

✔ A shuttle flight has been scheduled for Feb. 18, 1988.
✔ Five astronauts have been named to crew the next launch.
✔ Shuttle main engines are being improved.
✔ Boosters, which caused the 1986 disaster, are redesigned.
✔ A team is looking over all critical hardware and maintenance.
✔ A quality control and safety team is in on the job.
✔ Improved weather forecasting is set up for landing sites.
✔ Safety at landing sites is being improved.
✔ A way to escape from a disabled shuttle is planned.
✔ Top NASA management has been overhauled.
✔ A new administrator is on the job.
✔ A new deputy administrator is working.
✔ New management is running the shuttle program.
✔ New directors are controlling the space station program.
✔ Three major shuttle field centers have new leadership.
✔ Several astronauts have been assigned to top positions.
✔ Shuttle management has been concentrated in Washington
✔ Space station authority has been centered in Washington.
✔ New associate administrator posts have been created.
✔ Plans combine shuttle and space station into one project.

Shuttle Managers Changed

NASA moved space shuttle management in November 1986 from Houston to Washington, DC.

Arnold Aldrich, who had been manager of the shuttle program at Johnson Space Center, Houston, was moved to NASA headquarters at Washington.

Astronaut Robert Crippen, who flew on four shuttle flights, is Aldrich's deputy for shuttle operations. That includes preparing the spaceplanes for flight and overseeing launch decisions.

Richard Kohrs, shuttle deputy manager at Houston, is deputy to Aldrich at Houston. He watches over planning, scheduling, engineering and daily operations of shuttles, cargo and ground equipment.

NASA said returning strong control to Washington sets up a well-liked kind of management for the shuttles similar to that which ran the Apollo moon landings. Crippen and Kohrs report directly to Aldrich at Washington. NASA had been criticized for not making clear whether high officials at a center such as Johnson Space Center should report to the center director or to shuttle managers at Washington. Aldrich takes over some duties of Richard H. Truly, associate administrator for space flight.

Astronaut Is Safety Watchdog

USAF Col. Frederick D. Gregory, who flew the shuttle to space in 1985 and has flown 40 different kinds of airplanes and helicopters, is overseeing the well-being of men and equipment from his new post in NASA's safety office.

The 45-year-old pilot has been an astronaut since 1978. He works with various government agencies and private industry to set policies which ensure safety of flight crews.

Throttle Down To 104 Percent

A U.S. House of Representatives committee decided in October 1986 that Congress and the White House must share blame, along with NASA and private industry, for the *Challenger* disaster in January 1986.

It asked NASA to hold shuttle engine throttles down to no more than 104 percent of rated booster power, rather than the 109 percent used in the past, because of inadequate safety margin. House committee members said they thought all involved had been pushing too hard for NASA accomplishments and starving NASA of necessary funds.

Shuttle Engine Fired

Following the January 1986 shuttle *Challenger* disaster, NASA test fired a shuttle engine on June 25, 1986. The firing on June 25 lasted less than two seconds in the National Space Technology Laboratories at Bay St. Louis, Missouri. It checked the rocket test stand to see if it was ready for a four-minute blast which came in July 1986. The Bay St. Louis lab has been ground-testing all shuttle engines since 1975.

New Solid Rocket Tester

NASA is building a second stand in Utah to hold the redesigned space shuttle solid rocket motor (SRM). The new stand, designed to reveal stress on the SRM during firings, will hold the rocket in a horizontal position during testing.

A government panel approved the horizontal test firings even though they don't exactly simulate the forces on the rocket which is vertical in an actual spaceflight blast off. The panel decided not to spend an extra $20 million and a year's time to switch to vertical tests.

The panel told NASA to design different boosters for future backup in case the O-ring fix doesn't work. NASA's plan for redesigning the booster rocket features a change to the infamous O-ring seal. The new design uses a third O-ring and a mechanical grip to prevent the leak of hot gas which led to the January 28, 1986, explosion of the *Challenger* shuttle killing seven astronauts.

Shuttle Tiles

NASA is checking heat-insulating tiles on the three remaining space-shuttle craft. At least 26 were found loose in 1986 on the belly of the shuttle *Discovery*.

The problem tiles were glued to unusually-thin felt pads placed on *Discovery* after an August 1985 flight. Most tiles are bonded to pads about twice as thick as those found with poor connections. The badly-bonded tiles were not flown to space.

All tiles on *Columbia, Atlantis* and *Discovery* are being inspected. A total of 222 tiles will be replaced.

The fourth shuttle, *Challenger,* was destroyed in a launch explosion Jan. 28, 1986. The next shuttle flight probably will be late in 1988.

Challenger 7 Mural At Capitol

A colorful painting of the seven astronauts lost in the January 28, 1986, *Challenger* accident has been unveiled in a corridor of the U.S. Senate in Washington, DC.

The mural is in the Capitol's Brumidi corridor, named for an Italian immigrant who decorated the Senate in a florid 19th Century style. Blank ovals, left to record future U.S. history, line the brightly-painted corridor. Across the corridor from the *Challenger* mural is a 1975 painting reflecting the landing of the *Apollo* moonship in 1969.

The *Challenger* mural was painted by Temple University professor Charles Schmidt.

FBI Investigates Company

The Federal Bureau of Investigation is investigating Morton Thiokol Inc., manufacturer of the U.S. space shuttle solid booster rockets involved in the 1986 *Challenger* disaster.

The Chicago firm makes the boosters in Utah. Employees reportedly are giving evidence in the criminal investigation.

Meanwhile, retired Morton Thiokol engineer Roger Boisjoly is suing the manufacturer on behalf of the U.S. government for $4 billion, saying the company filed false payment claims with NASA.

Shuttle *Challenger* exploded during lift off January 28, 1986, killing seven astronauts. All shuttles have been grounded since.

The next flight may not come until late 1988. Officially, it has been scheduled for February 18, 1988. Reports indicate NASA won't be able to meet the February schedule as engineers test booster rockets and make necessary changes.

Discovery is the next spacecraft in line for flight. There were press reports of rumors among astronauts that the *Discovery* flight would be delayed until after the 1988 presidential elections so an undesirable setback wouldn't embarrass the Reagan Administration.

Whenever shuttle flights resume, the public will see more flights postponed during countdowns. New safety rules will be less flexible. More scheduled launches will be postponed as problems occur.

Don't Rely Solely On Shuttle

NASA Advisory Council in March 1987 told the U.S. space agency to quit putting all its eggs in one basket, stop relying only on the shuttle fleet for space launches, and use expendable rockets.

The council recommended hauling as much cargo as possible with unmanned rockets. The shuttle would be used only where manned launches were a necessity.

The advisors told NASA that America's space leadership has been damaged. If expendable rockets aren't used, it may be impossible to regain the leading role.

Expendable launch vehicles (ELVs) were phased out after a 1970s decision to rely solely on shuttles. The 1986 *Challenger* disaster grounded all shuttles and sent the U.S. space program into a tailspin. NASA has only a couple of large boosters in stock. Important projects have been delayed and funds have been diverted to repairs on the shuttle. The 25-member Advisory Council, the agency's most important external board, wants NASA and the Reagan Administration to ask Congress for extra money for unmanned rockets. ELVs could get several important satellites to space two to four years earlier than waiting for shuttle rides.

NASA's independent Aerospace Safety Advisory Panel made the same recommendations.

NASA Finishes Burial For Now

U.S. space agency workers have finished putting most of the shuttle *Challenger* wreckage in storage in old *Minuteman* ICBM underground complexes at Cape Canaveral, Florida.

Giant concrete caps were lowered onto the abandoned missile silos in February 1987.

The spaceplane and its payload are in four underground buildings. Pieces of the external fuel tank and solid-fuel booster rockets are in two 78-foot-deep silos.

Challenger exploded during take off January 28, 1986, killing seven astronauts.

Parts of the right-side booster which ruptured in the explosion will be kept elsewhere for a year for study by engineers.

The Smithsonian Institution's National Air and Space Museum wanted to display pieces of wreckage in Washington but NASA didn't think it proper.

Secret *Challenger* Photos

NASA decided not to release photos of the shuttle *Challenger* crew cabin. Officials said it would be an invasion of privacy of the dead crew and their families.

The wreckage, from the January 1986 explosion, is stored in a *Minuteman* missile silo at Cape Canaveral Air Force Station. Photos and videotapes, which reportedly include shots of astronaut personal effects strewn across the ocean floor, will be shown only to NASA workers.

Photos of other parts of the wrecked shuttle are available and one photo of crew cabin debris was released in 1986 before the policy changed. Reporters toured the Kennedy Space Center hangar where *Challenger* wreckage was held before burial.

Challenger Point

A 14,081-ft. mountain peak near Kit Carson Peak in the Sangre de Cristo range in southcentral Colorado will be named *Challenger*

Point to honor astronauts killed in the 1986 shuttle explosion.

The U.S. Board on Geographical Names voted for the name which must be approved by the Interior Secretary. Dennis Williams, Colorado Springs electrical engineer who suggested the idea, said mountain peaks symbolize the spirit of adventure.

Some residents opposed the name. County commissioners voted against it. The sheriff said residents were afraid it would encourage people to climb and fall off.

Ten have died in eleven years from climbing accidents and small plane crashes near the point. Supporters included a former state governor and the Colorado Mountain Club.

Cosmonauts Carry Pictures

USSR cosmonauts Vladimir Soloyov and Leonid Kizim took a photograph of the seven U.S. shuttle *Challenger* astronauts up to the *Mir* space station in 1986.

The cosmonauts, launched six weeks after the *Challenger* exploded in January 1986, "wanted those seven brave astronauts to go to outer space." While visiting the U.S., Kizim and Soloyov said they had paid tribute in space to the dead U.S. astronauts.

They spent four months in the *Mir* space station. Soloyov has spent more days in space than any other person. He has logged a total of 375 days in orbit during three flights. *Mir* is Russian for *Peace.*

The spacemen accompanied a teenage Young Cosmonauts group from the USSR. They toured the U.S. launch facility at Cape Canaveral, FL, as well as Washington, Disney World, and NASA space centers at Huntsville, AL, and Houston, TX. Earlier, Young Astronaut teens from the USA toured the USSR.

Other O-Rings

NASA isn't the only place where O-rings cause concern. There are thousands of O-rings in nuclear power plants across the country protecting against radiation leaks. More than 60 reportedly have failed dating back to 1975. No breakdown led to a radioactive leak.

Some scientists are concerned that the O-rings, designed to withstand 4,000 degrees Fahrenheit temperature, will hold up at the 13,000-degree temperature of a nuclear accident. The O-rings are in cooling systems preventing cores from overheating. They also are in shock absorbers preventing earthquake damage.

Next Shuttle Flight In 1988

NASA has moved the official launch date to return its shuttle fleet to active duty to June 1988. However, that date may slip to later in 1988 before *Discovery* lifts off from Cape Canaveral.

Whenever it is set, the public will see more flights postponed during countdowns. New safety rules will be less liberal, less flexible. More scheduled launches will be scrubbed than before.

Some 400 NASA workers have thumbed through four million pages of documents about old launch procedures to find which will be safest for the future. More than 2,700 launch site hazards have been checked. And they are keeping a close eye on shuttle engine designs.

One item getting new attention is the fuel line from external tank to shuttle engine. There is a possibility a 17-in. valve where hydrogen and oxygen are connected could close, causing an explosion. Some 18,000 gallons of oxygen and 49,000 gallons of hydrogen flow through the valves each minute. The pressure of the flow should hold them firmly open. In recent tests, they looked like they were fluttering. Welds in hard-to-see areas also are causing concern. Before any shuttle flights, the solid-rocket booster must pass tests.

The National Research Council has recommended that NASA:

✔ set up a ranking of critical items in order of which is most likely to fail;

✔ compare actual engineering changes with those called for to see they are made;

✔ add computer analysis to enhance seat-of-the-pants decisions now made by engineers;

✔ focus attention of top NASA management on the most critical items when deciding whether or not to call off a flight.

The particular booster rocket which caused the *Challenger* explosion was used with special permission from an official at Marshall Space Flight Center, Huntsville, AL.

The National Research Council is an independent study panel headed by retired Air Force Gen. Alton D. Slay. The group's report praised NASA for taking a second look at hundreds of procedures and items which might cause a disaster.

Frederick H. Hauck, commanding the *Discovery* crew, said in January 1987 design changes and safety requirements can't be met by early 1988 and launch would slip to a later date.

Shuttle Commerce

Before the *Challenger* explosion, NASA had contracts with 44 private firms to launch satellites via shuttle. Since the disaster, NASA has told 24 companies they can't ride in the shuttle until after 1994.

NASA will have to reimburse a total of $40 to $50 million for the 24 cancelled launches.

Among those bumped were 10 Hughes satellites, four RCA, and an American Satellite Co. craft. NASA still will launch one RCA and one Hughes satellite sometime in the next seven years. Some of the dislocated satellites have been switched to the European *Ariane* launcher.

Discovery Prepped

The space shuttle *Discovery* was moved from the huge vehicle assembly building at Cape Canaveral, Florida, to an orbiter processing facility in October 1986. It is being prepared for the next shuttle flight, sometime in 1988.

Astronauts on that first 1988 flight will be Frederick H. "Rick" Hauck, Air Force Lt. Col. Richard O. Covey, John M. "Mike" Lounge, George D. "Pinky" Nelson, and Marine Maj. David C. Hilmers.

NASA also rolled the *Atlantis* shuttlecraft up to its Cape Canaveral launch pad for ground tests by an astronaut crew: commander Robert L. Gibson, pilot Charles F. Bolden, and Franklin R. Chang-Diaz, Steven A. Hawley and George D. Nelson.

NASA also tested a new emergency escape system for astronauts caught in launch pad fires. Astronauts for that *Atlantis* on-the-ground exercise were commander Frank L. Culbertson; pilot Stephen S. Oswald; and astronauts Carl J. Meade, Kathryn C. Thornton, G. David Low, Pierre J. Thuot and Jerome Apt.

U.S. Shuttle Schedule

Challenger, destroyed in January 1986, was shuttle flight number 25. Flights 26 through 36 are set to take place from 1988 through 1990.

Three to five U.S. shuttle flights have been scheduled for 1988. Ten shuttle flights are scheduled for 1989. Here's the schedule of payloads for the next eleven flights:

Shuttle 26 The first 1988 flight will carry NASA's *Tracking and Data Relay Satellite TDRS-C* aloft for communicating data between shuttles and ground controllers. *TDRS-C* replaces a *TDRS* destroyed when shuttle *Challenger* exploded during shuttle flight 25. Many of NASA's future plans for space shuttles require a pair of satellites in orbit so *TDRS-D* will be carried to space in shuttle flight 29.

Shuttle 27 The second and third flights of 1988 will be secret Department of Defense missions. A secret military communications, missile warning or electronic intelligence-gathering satellite bound for a 22,300-mile-high stationary orbit will ride to space in flight number 27.

Shuttle 28 The third flight will carry a secret CIA spy

photography satellite.

Shuttle 29 The fourth shuttle flight, which may be early in 1989, will ferry NASA's *TDRS-D* satellite to orbit.

Shuttle 30 The fifth launch will be of immense importance to mankind. The 2.4-meter *Hubble Space Telescope* will be carried aloft to study deep space and the edge of the Universe for 20 years.

The *Hubble Telescope* was to have been placed in orbit by a shuttle in the summer of 1986.

It will be able to see stars so far away they sent out their light during the earliest time of the Universe. Shuttles will make service calls at the telescope during its 20-year life. Astronomers have called the *Hubble Telescope* "the greatest scientific instrument that man has ever built."

The *Hubble Space Telescope* passed a simulated 28-orbit ground test in March 1987 when five science instruments on the telescope in storage at Sunnyvale, California, were controlled remotely by telephone from Goddard Space Flight Center at Greenbelt, Maryland. NASA conducted the three-day test to reassure itself about the billion-dollar telescope.

Shuttle 31 A group of scientific ultraviolet telescopes known as *ASTRO-1* which will remain with the shuttle.

Shuttle 32 Another secret military satellite will be started toward stationary orbit at 22,300 miles from Earth.

Shuttle 33 An important exploration satellite, *Magellan,* will be aboard the third shuttle flight in 1989. *Magellan* will use a sharp radar eye to map the planet Venus. It will be the first new U.S. interplanetary venture in more than a decade.

Shuttle 34 This will be a secret military Star Wars weapons Spacelab flight.

Shuttle 35 A materials-science research payload plus two military navigation *Navstar* satellites will be ferried to space in the shuttle.

Shuttle 36 Another secret military satellite will head for stationary orbit at 22,300 miles from Earth.

Other flights. NASA wants to launch the *Galileo* spacecraft to Jupiter from a space shuttle in October or November 1989.

Eleven launches are planned for 1990. In July 1990, the shuttle will carry the British Ministry of Defence's *Skynet 4* military communications satellite to orbit.

NASA hopes to launch the European Space Agency's *Ulysses* craft, to study the Sun close up, from a shuttle in October 1990.

NASA will launch a communications and weather satellite for India in October 1990, three Space Industries Inc. satellites in 1992 and 1993, and three Geostar Corp. navigation satellites in the early 1990s.

The U.S. may send tv cameras and research instruments out to Mars in 1992. The USSR is expected to launch the first of a series of Mars explorer craft to the planet and the Martian moons in 1988.

NASA expects to start carrying space station construction materials to orbit in 1994.

Heavy military. From 1988 through 1994, more than 40 percent of shuttle flights will be secret Defense Department operations. Four of the ten 1989 flights will be Defense Department flights. Among payload items on two other 1989 flights will be navigation and positioning satellites, part of the military Global Positioning System (GPS). Four of eleven launches in 1990 will be secret military missions and three other shuttle flights that year will carry GPS satellites.

Navstar satellites are scheduled to tie up the shuttle for many trips to space after flight 36. The Air Force may buy new *Delta 2* expendable rockets for the *Navstar* satellites to free shuttles to ferry other payloads to space in the 1990s.

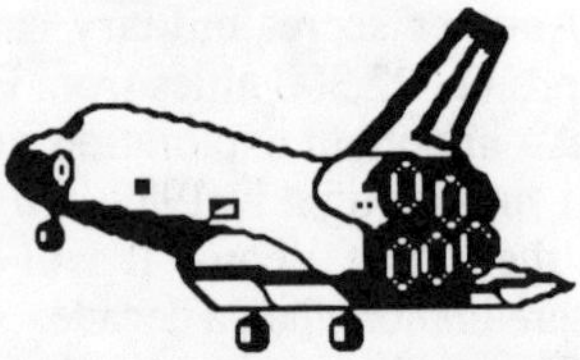

Next Shuttle Crew

Judging by their past space exploits, NASA will be sending its five best astronauts to space in the planned 1988 *Discovery* launch.

★Frederick H. "Rick" Hauck, 45, in charge of a November 1984 shuttle flight which brought two communications satellites down from orbit, will command the first crew to fly since the January 1986 *Challenger* disaster.

Hauck, a nuclear engineer and former combat pilot, has been NASA external relations associate administrator. He has spent a total of 339 hours in space on two flights. Hauck was pilot on a 1983 shuttle flight.

His 1984 flight was the first salvage mission. The two satellites plucked from space were Westar and Palapa. They had been ferried to space in an earlier shuttle but refused to fly properly on their own. Hauck's crew put them in the shuttle cargo bay and brought them down to Earth to be fixed. They have been sold and are expected to be taken back up to orbit.

★Air Force Lt. Col. Richard O. Covey, 40, pilot on a 1985 shuttle *Discovery* flight when two spacewalkers repaired the *Syncom* communications satellite, will pilot *Discovery* in February 1988.

Covey as a fighter pilot flew 339 combat missions in Southeast Asia. As on-the-ground "capcom" capsule communicator, he was the last person to talk to the *Challenger* crew before the shuttle blew up in January 1986.

★John M. "Mike" Lounge, 40, a mission specialist on that 1985 *Discovery* flight with Covey, also will be mission specialist on the February 1988 flight. Lounge, a Navy combat pilot, holds a rare masters degree in astrogeophysics.

★George D. "Pinky" Nelson, 36, who took an untethered spacewalk in 1984 using a jet backpack, will be a mission specialist.

Nelson, an astrophysicist, tried unsuccessfully during his untethered spacewalk to capture the broken *Solar Max* satellite by hand. The gadget he was using to capture the satellite failed but *Max* later was grabbed by the shuttle's giant robot arm. Nelson and another astronaut repaired *Max* in the shuttle bay and pushed it overboard, back in orbit.

Nelson walked in space untethered with his jet backpack twice for a total of nine hours. He also went up in January 1986 aboard the shuttle *Columbia,* last to make it to space.

★Marine Maj. David C. Hilmers, 36, mission specialist on a secret military shuttle *Atlantis* flight in October 1985, again will be mission specialist. He was honored as an athlete and scholar and is an engineer and pilot for the Marine Corps.

The *Discovery* launch in 1988 will be the 26th shuttle flight. It will be the first no-rookie shuttle crew—only astronauts who have been in space before. Seven crew members were killed when shuttle *Challenger* exploded a minute after launch Jan. 28, 1986. There have been no shuttle launches since.

Shuttle Escape Plan

If they have five to ten minutes to set it up after all engines fail, U.S. shuttle astronauts may have an escape system when they fly again in 1988.

If all engines fail when the shuttle is in a controlled glide at about 20,000 feet altitude, an entire crew might be able to escape by the time the spaceplane drops to 10,000 feet or a bit lower.

The system, if available back in January 1986, would not have saved the seven *Challenger* astronauts whose craft started exploding seconds after liftoff.

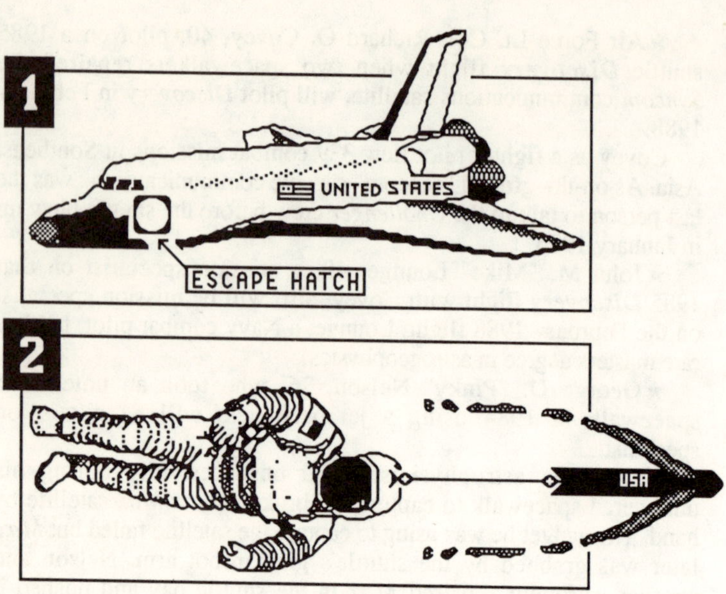

IN AN EMERGENCY, the shuttle entry door (1) would be blown open, becoming an escape hatch. An astronaut would lie down on a ramp with his head pointed toward the door. A small rocket with a 12-foot cord would be fired, pulling the astronaut (2) out, clear of the left wing and away from the shuttle.

The next shuttle flight is likely late in 1988. The escape mechanism under consideration would be useful if boost was lost later in a flight to orbit. Bailing out might be preferable to ditching inside the shuttle in an ocean.

In an emergency, astronauts inside the shuttle would use the so-called powered extraction system to blow off the shuttle's entry door, turning it into an escape hatch. They would attach a stretcher or tray to the hatch. An astronaut, wearing a parachute, would lie down on the stretcher and hook a 12-foot cord between his body harness and a small rocket pointed out the door. The rocket would be fired, tugging the astronaut at ten times the force of gravity.

NASA has not given final approval to the system but engineers are working on it. It is not clear whether the system will be available for the 1988 shuttle flight or if that flight would be delayed for completion of the escape system. Astronauts say they want an escape system.

The proposed system would add 600 pounds weight. NASA had

decided against escape systems for the shuttle years ago. It had been thought they were too heavy and cost too much. A loaded orbiter weighs about 120 tons.

If a shuttle lost power near space and could not return to land in Florida, it might have to ditch in an ocean. Astronauts might not survive a crash into the ocean at 230 miles per hour. If they bailed out, they probably would have to wait in the water from two to 24 hours for rescue. They would have a radio beacon, survival gear and life rafts. Engineers say the time in the water would be more dangerous than the rocket-powered escape from a shuttle.

Studies also are underway to find more sophisticated escape systems to save astronauts from Challenger-type accidents in future generations of shuttles.

Fishing Reel Satellite

An Italian satellite will be cast on a 12.5-mile tether from a large reel on the U.S. shuttle *Columbia* to fish for electrical power in space in the fall of 1990.

Italian scientists will build the satellite for the *Tethered Satellite System (TSS-1)*. Americans will build the reel and tether, and fly it all to space in *Columbia*.

Scientists want to find out what happens when a retractable cable and satellite are dragged through space by a shuttle. They plan to tether microgravity laboratories to the space station eight years from now.

Experiments will measure electricity generated when the tether passes through ion clouds in Earth's upper atmosphere. Shuttle altitude ranges from 120 to 350 miles high.

Experimenters believe the shuttle, tether, satellite, ions, and Earth's magnetic field could interact to produce up to 5000 volts of electricity at two amps current yielding ten kilowatts of electrical power. That current flows in the tether creating radio signals and static.

Such signals and static could affect astronomers on the space station as they try to hear weak radio noise from deep space galaxies, quasars, pulsars and the like. Also, the electrical power generated by the tethered satellite might be captured to turn on lights inside the station when it is permanently manned after 1994.

Original Space Shuttle Idea

The U.S. Air Force came up with the first concept for a space shuttle back in 1962.

They had their B-52 bombers and an experimental rocket plane known as X-15. The idea was to bolt an unmanned X-15 to the top of a B-52 and fly it as high as the plane could go. From that altitude, near space, they would launch the unmanned rocket plane. The X-15 would drop off a satellite in orbit and fly back down to Earth.

It never flew. The idea was dumped in 1965.

Shuttle 2

The current generation of space shuttles eventually will wear out. NASA is looking into a second generation shuttle to replace them in the early 2000s.

The next orbiter might be smaller but cheaper and more powerful.

NASA Doublespeak

NASA won the 1986 Doublespeak Award from a national panel of English teachers for this explanation:

"The normal process during the countdown is that the countdown proceeds, assuming we are in a go posture, and at various points during the countdown we tag up the operational loops and face to face in the firing room to ascertain the facts that project elements that are monitoring the data and that are understanding the situation as we proceed are still in the go direction."

The award was granted at the 76th annual conference of the National Council of Teachers of English in November 1986.

Running close for the doublespeak trophy were bits and pieces of military jargon including aluminum transfer case (a temporary coffin), manually powered fastener driving impact device (a hammer), emergency exit light (a flashlight), hexiform rotatable surface compression unit (a steel nut), and a frame supported tension structure (a tent).

NASA Space Camp

NASA's Space Camp, a one-week camp session for students who have completed 6th, 7th or 8th grades, shows kids what being an astronaut would be like.

Information is available from United States Space Camp, The Space and Rocket Center, Tranquility Base, Huntsville, Alabama 35807.

Satellites

17,304 Objects In Orbit

There have been 17,304 objects sent to space by various countries since the dawn of the Space Age in 1957.

Any man-made object in space—a satellite payload, a used rocket, pieces of debris after an explosion, even a camera lens—is included in the total.

Useful payloads over the years have totaled 3,559. The other 13,745 objects have been space junk.

The top five countries, not including international organizations, which had the largest number of their own useful payloads in space on December 31, 1986 were the USSR which had 978; the USA

SCOREBOARD OF SATELLITES AND SPACE OBJECTS
AS OF DECEMBER 31, 1986

	OBJECTS IN ORBIT			NO LONGER IN ORBIT		
COUNTRY	PAY	DEB	TOT	PAY	DEB	TOT
AUSTRALIA	3	2	5	1	0	1
BRAZIL	2	0	2	0	0	0
CANADA	14	0	14	0	0	0
CZECH	0	0	0	1	0	1
ESA	15	99	114	1	5	6
ESRO	0	0	0	7	3	10
FRANCE	14	21	35	7	49	56
FRANCE/FRG	2	0	2	0	0	0
FRG	5	1	6	3	5	8
INDIA	7	4	11	2	3	5
INDONESIA	3	0	3	1	0	1
ITSO	35	0	35	0	0	0
ITALY	1	0	1	4	0	4
JAPAN	31	40	71	5	26	31
MEXICO	2	1	3	0	0	0
NATO	6	0	6	0	0	0
NETHERLANDS	0	0	0	1	3	4
PRC	4	3	7	15	39	54
SAUDI ARABIA	2	0	2	0	0	0
SPAIN	1	0	1	0	0	0
SWEDEN	1	0	1	0	0	0
UK	9	1	10	6	4	10
USA	520	2525	3045	567	2070	2637
USSR	978	1885	2863	1283	6956	8239
TOTALS	1655	4582	6237	1904	9163	11067
GRAND TOTAL						17304

PAY: PAYLOAD DEB: DEBRIS TOT: TOTAL

with 520; Japan with 31; and France and Canada, both of which had 14.

Of course, both France and Canada have contributed heavily to multinational satellites. The European Space Agency had 15 in orbit but ESA is a multinational cooperative.

The countries with least useful payloads in orbit were Italy, Spain and Sweden with one each. Two other countries, Czechoslovakia and The Netherlands, each used to have one but had no satellites in orbit at the end of 1986.

Long-Life Satellites

Comsat has found a way to extend the useful life of a satellite already in orbit without flying out to refuel it.

The communications satellite company has developed what it calls the *Comsat maneuver* to make a satellite use less fuel to hold its position in the sky.

A stationary satellite is maneuvered to an exact spot in space 22,000 miles above Earth. To maintain good contact with the antenna of a ground station, the satellite must remain in that exact spot in the sky. As the satellite orbits the Earth at the same speed our globe rotates, the satellite appears stationary to those looking up from the ground. The signal from such a stationary satellite always beams down on the same area of the Earth. The surface area covered by that beam is called a footprint.

There's one small problem. A satellite will drift around the sky. A satellite only drifts a small distance. The drift is predictable, mostly back and forth in a north-south direction.

Each satellite has a small amount of fuel to power tiny rockets which, on command, push the satellite back into its assigned spot. Over several years, a satellite can use a lot of fuel to hold its position. If the north-south drift is not controlled, the radio footprint also moves north and south on the surface of the Earth as the satellite drifts. Sometimes, ground stations lose contact with a satellite.

The Comsat maneuver changes the angle of the satellite as it looks down on the Earth. It still moves in the north-south drift but the radio footprint stays over the same Earth surface area. Ground stations maintain constant contact with the satellite.

Looking up from the ground, the satellite still drifts north and south like clockwork. But ground antennas track that small motion. Contact is not lost. Less fuel is used from the satellite to keep it on station. The result is an extension of the useful life of a satellite.

Comsat's *Comstar* satellite had been using 37 pounds of fuel a year to keep itself on station. After the Comsat maneuver, it uses only three pounds of fuel a year.

Heat Pipes Cool Satellites

Engineers will be building water-filled heat pipes to cool electronic gear on board future spacecraft.

The pipes are sealed tubes filled with water, or sometimes methanol or ammonia. When an end of a pipe is placed near heat, the fluid in the pipe vaporizes and floats off to the other end where it condenses back into liquid and gives off its heat. The condensed fluid is sucked up by a wick and carried by capillary action back down to the hot end of the pipe where the action starts all over.

French Debris

The amount of space junk in orbit around Earth went up dramatically in November 1986 when an European Space Agency *Ariane* rocket exploded in orbit.

The third stage of the French rocket had been in orbit nine months when it blew apart, creating a cloud of debris which might threaten satellites.

Radar on the ground shows 200 pieces of debris of one-half-inch diameter or larger. Thousands of tiny particles, too small for radar to see, may be floating in the cloud of debris. If a tiny particle collided with it, a satellite or manned spacecraft could be damaged or destroyed. Many satellites are in orbits which pass through the area where the debris is located.

The rocket was launched Feb. 22, 1986, ferrying the *Spot* satellite to orbit. *Spot* makes photos of Earth for commercial use.

The third stage of *Ariane* used liquid oxygen and liquid hydrogen propellants, the same explosive fuel as the U.S. space shuttle. The harsh space environment eventually exploded small left-over quantities of oxygen and hydrogen in *Ariane*.

Another *Ariane* rocket, launched May 30, 1986, was destroyed by ground controllers when its third stage rocket failed. That problem put the European Space Agency launch program on hold. Two other *Ariane* third stages have failed at launch.

The problem is not unique to *Ariane*. There already was plenty of debris in the same area of space from explosions of second stages from seven U.S. *Delta* rocket flights. They exploded many years ago after long periods in space. *Delta* was modified to prevent such explosions.

Hot Start In 1987 Deep Freeze

The USSR launched nearly a dozen satellites during the frigid weeks of January and February, 1987.

The Russians expected to send some 100 satellites to orbit in 1987. The number compares with six launches planned by NASA for 1987.

Most of the USSR satellites will be named *Cosmos*. Numbers designating different *Cosmos* satellites are approaching 2000. Names of other Russian satellites may include *Raduga, Molniya, Meteor, Gorizont* and *Radiosputnik*. At least two amateur radio satellites in the *Radiosputnik (RS)* series are expected to be carried to orbit.

Meteor. The fifteenth *Meteor 2* weather satellite was sent to a 600-mile-high orbit from the Plesetsk northern cosmodrome on

January 5 despite subzero temperatures which disrupted central Europe and Asia.

Unofficial radio listeners in England reported the new *Meteor's* radio signals were interfering with transmissions from an older *Meteor* in orbit and a satellite known as *Cosmos 1766* which uses radar for ocean surveillance. All three satellites apparently were transmitting on a frequency of 134.7 MHz. On January 14 the USSR changed the new satellite to a frequency of 137.85 MHz. The older *Meteor* may be stuck on the air on 134.7 MHz, interfering with the military ocean radar satellite.

Molniya. A *Molniya 3* communications satellite was placed in an elliptical orbit January 23.

Cosmos. An ultramodern photography reconnaissance satellite, *Cosmos 1811*, was launched from the Baikonur Cosmodrome at Tyuratam in frigid central Asia on January 9. It ranges out to 228 miles.

Cosmos 1812, an electronic intelligence satellite, or radio ear in the sky, was launched from the Plesetsk cosmodrome on January 14 into an orbit 400 miles high.

Cosmos 1813, another photo spy satellite, was launched January 15. It was parked for a short time in an orbit ranging from 130 to 240 miles out. Later it was moved to a 250-mile-high orbit for a better view of the United States and Western Europe. The military reconnaissance satellite flew two weeks and was supposed to drop its film to a Russian recovery team. But, when it did not respond to re-entry commands, USSR controllers exploded it in space to prevent danger from an uncontrolled impact of the recovery capsule which weighed several tons. The explosion, which blew the spacecraft into more than 100 pieces of space debris out to 350 miles, also prevented recovery by others.

Cosmos 1814, was launched January 21 into a 500-mile-high orbit. The communications satellite receives signals when over one ground station and retains those messages until commanded to send them down by a different station elsewhere.

Cosmos 1815 is a military satellite launched January 22 from the Volgograd cosmodrome at Kapustin Yar into an orbit ranging out beyond 300 miles.

Cosmos 1816 is a navigation satellite, 600 miles in space, launched January 29.

Cosmos 1817, possibly a communications satellite on its way to stationary orbit at 22,000 miles out, was destroyed when the *Proton* launcher's fourth stage failed to fire from its 150 mile high parking orbit after lift-off January 30.

Cosmos 1818, launched February 1, is in a unusual 500 mile high orbit. It may be a communications satellite or an ocean surveillance satellite.

Progress 27. In 1986 the USSR launched a second-generation space station, *Mir*. An unmanned Earth-to-space freighter, *Progress 27*, loaded with food, fuel and supplies for the space station, was launched in subzero temperatures January 16, 1987 from Tyuratam.

It parked in an orbit ranging from 114 miles out to 164 miles. Two days later its engine was used to power it out to 200 miles for a rendezvous and docking with the *Mir* space station. *Progress 27* then pushed the combined freighter and space station to a circular orbit about 215 miles above Earth.

Soyuz. On February 6, 1987, cosmonauts Yuri Romanenko and Alexander Laveikin were launched from Baikonur Cosmodrome in a new-style *Soyuz TM-2* capsule to board *Mir*. After docking *Soyuz* at a port on *Mir*, the cosmonauts unloaded provisions from *Progress* into *Mir*.

The first addition to *Mir* in 1987 was an x-ray observatory including electronics from the European Space Agency sent up in April. The Soviet Academy of Sciences, the European Space Agency, and West Germany each supplied special spectrometers. Great Britain and the Netherlands supplied a telescope for electronic images.

A 42,000 lb. 43-foot-long duplicate of *Mir* will be flown to orbit as an add-on module to double the size of the space station. The Russians are said also to have built a small module for connection to the space station. It is not known whether another small module as well as a full-sized *Mir* module will be sent to space in 1987.

USSR Science Craft Coming

The Russians are building ten major scientific platforms and robot observatories to send to Earth orbit plus other craft to travel to Mars in the next five years:

Aktivny-1K is a satellite to be launched into Earth orbit in 1987 to study the magnetosphere surrounding the globe. A small satellite, *Magion*, will be released from the main craft as part of the study.

Biosat will carry two monkeys, several rats and other small animals to orbit in 1987 in a biological life-sciences flight to see the effects of weightlessness on their bodies.

Gamma-1 is a large USSR-French astrophysics package with a 3,300-lb. telescope to study gamma rays and x-rays from deep space. The satellite is built in a modified *Soyuz* and *Progress* spacecraft shell.

Roentgen is a 40,000-lb. astrophysics telescope to be launched on a *Proton* rocket to dock with *Mir*. It is not clear if this is the same as *Quantum* but *Roentgen* has three West European telescopes and one USSR instrument.

Earth resources module is a 40,000-lb. photography module to dock with *Mir* in 1987. Photographing mineral and fuel resources on Earth's surface from space offers large economic benefits to the USSR.

Granat in 1988 will be a French-Russian 2,200-lb. telescope to study x-ray stars in deep space. It will follow the work of *Gamma-1* and *Roentgen.* It will be constructed in a new, larger general-purpose *Astron* spacecraft. *Astron* has been planned to handle large telescopes and other space-science projects until after the year 2000.

Interbol is yet another USSR-French project with four spacecraft to be sent to orbit in 1988 and 1989. Two *Progress* and two *Magion* spacecraft will study the magnetosphere following the *Aktivny-1K* and *Magion* flights in 1987.

Aelita will be a 14,000-lb. *Progress* spacecraft sent aloft in 1990 to study very-short-wavelength signals from deep space. It also is a cooperative project with France.

Rascas will be a radiotelescope in space with a 100-ft. diameter. The Russians already tested a 33-ft. antenna known as *KRT-10,* released from the *Salyut 6* space station in 1979. The satellite, to go to Earth orbit between 1988 and 1990, will use the technique of interferometry, working in conjunction with telescopes on the ground in the Crimea to study signals received from deep space.

Radio Astron, launched between 1990 and 1992, will use a 33-ft. antenna to receive signals from deep space. The European Space Agency will take part in the experiments.

Atlantic Ocean's Spring Bloom

A weather satellite peering down at the ocean since 1978 charted all of the plankton near the surface of the North Atlantic.

Plants on land and in the ocean take carbon dioxide from the atmosphere and the oceans and change it into plant tissue in the important process, photosynthesis. NASA's unmanned *NIMBUS 7* weather satellite, in orbit since October 1978, stared at sunlight reflecting from a 500-mile-wide swath of ocean every time it flew overhead, noting the color of the surface of the sea.

Information was sent by radio to scientists on the ground. Many such images were combined by the scientists to make an ocean-wide color picture of how the microscopic plant life, phytoplankton, looks to the satellite.

The most striking feature is *spring bloom,* a big concentration of plankton extending all the way across the North Atlantic into the North Sea.

Scientists used mathematical equations on the satellite data to compute how much chlorophyll is in the water. Chlorophyll is a pigment giving plants such as phytoplankton a photosynthesis ability.

NIMBUS sees five colors of light. It was built to scan coastal waters. The new picture shows land and ocean plants on the edges of continents bordering the North Atlantic.

Land vegetation was recorded by a sharp-eyed *NOAA 6* satellite, in orbit since June 1979. The Sahara Desert, tropical rainforests of South America and the spring greening of North American fields and forests are major features seen by *NOAA 6*.

The *NIMBUS* ocean scanner died in 1986 but NASA plans an improved land and ocean eye for science aboard the U.S. space station in the 1990s.

France, U.S. To Study Ocean

CNES, France's space agency, and NASA, the U.S. space agency, jointly will launch a *Topex/Poseidon* satellite in 1991 to map surface features of oceans around the globe.

The satellite will be built in Maryland. CNES, the Centre National d'Etude Spatiale, will launch *Topex/Poseidon* to orbit with one of its *Ariane* rockets from South America. NASA and CNES both will provide sensors for the satellite. They will share data received from it.

Topex/Poseidon will measure the sea level of Earth's oceans at least until 1994. Scientists expect to determine the general circulation pattern of the oceans, see surface features, and find a better understanding of how oceans affect climates.

Rift Valley Fever

Mosquitos carried a virus from the Rift Valley on the eastern side of Africa across the continent to Egypt in 1977 killing 2,000 people with its deadly fever.

Now a Maryland research laboratory has found a way to track the path of the disease by satellite.

The fever killed most of the sheep in Egypt. Some 200,000 humans caught the fever while slaughtering sheep.

The Rift Valley stretches from Mozambique to the Red Sea. The fever was first diagnosed there in 1968.

Researchers today use satellite photos of land conditions across Africa to predict spread of the fever virus. They map vegetation density which gives clues to mosquito movements.

The 1977 outbreak was stemmed by animal vaccine from South Africa and human vaccine from Fort Detrick, Maryland.

Japan Launches Two

Japan continued a steady march into the space age in February 1987 by successfully launching *Astro-C* and *MOS-1* satellites.

Astro-C left Kagoshima Space Center February 5 on a science trek in Earth orbit to observe deep space x-ray objects. The Maritime Observation Satellite *MOS-1* was blasted to space February 18 aboard a Mitsubishi N-2 rocket.

Equipment to look at x-ray stars aboard *Astro-C* includes instruments from the U.S. Los Alamos (NM) National Laboratories and the United Kingdom's University of Leicester and Rutherford Appleton Laboratory.

MOS-1 will provide data to 60 Japanese organizations as well as 16 other countries. The N-2 booster is similar to the U.S. Delta. The Japanese are expected to improve their lift-to-orbit capability by turning the N-2 into a two-stage rocket. The two-stage H-1 already has been tested successfully. The last N-2 in Japan's inventory will be used to launch *MOS-2*, a companion to *MOS-1*.

The 1,628-lb. *MOS-1*, in a 564-mile-high orbit, is only the latest good news in a string of successful launches from Japan.

Ajisai Mirrorball Visible In Sky

A glowing seven-foot sphere called *Ajisai*, launched into orbit in August 1986 aboard Japan's H-1 rocket, is visible to the naked eye on the ground just before sunrise and right after sunset.

Ajisai means Hydrangea, a flower which Japanese builders say the satellite resembles. *Ajisai* was called Experimental Geodetic Payload or EGP before launch. Japan traditionally names its satellites for flowers after launch. *Ajisai*, pronounced "ah-high-sy," is a totally passive mirrorball like a big rotating disco flasher in the sky. It has no electronics aboard.

The laser ranging and experimental sky position satellite, studded with hundreds of reflectors, was the primary payload carried aloft by the H-1 booster which also lofted *Fuji*, an amateur radio satellite. When the sky is clear it is possible to see *Ajisai* from a quarter hour to two hours after local sunset and before sunrise. It will look like a bright star—similar to *Polaris*, for instance—and could take up to 18 minutes to cross the sky above you.

The satellite spins at 40 revolutions per minute, reflecting two sunlight flashes per second. It has an apparent brightness of first magnitude to fourth magnitude, typically looking as bright as a second magnitude star. Your eye can see fifth or sixth magnitude stars without help so you should be able to see *Ajisai* easily when it is overhead.

Launched by the Japan Hydrographic Department, the satellite provides a target for precise calculation of locations on Earth of geodetic stations. Stations in Maryland, California, Australia,

England, Switzerland and elsewhere around the globe have been sending up laser beams which bounce off the reflectors and return to Earth. The time it takes a beam to go up and back tells scientists how far away *Ajisai* is—its range. Photography shows its position against the background star field.

Viewers looking at 1510-pound *Ajisai* through binoculars sometimes see a faint object leading along the orbital path, about three seconds ahead. That is the second stage of the H-1 rocket which placed *Ajisai* and *Fuji* in orbit August 12, 1986. *Fuji*, flying near the two visible objects, is too faint to see. *Fuji* means wisteria.

The H-1 vehicle was fired into space from Japan's Tanegashima Launch Site on an island off the coast of South Kyushu, Japan. The surface of *Ajisai* is 85 percent covered by 318 convex mirrors which reflect sunlight. The periphery of the satellite is lined with 1,436 cubes of polished fused silica glass. The cubes are in 120 clusters of 12 cubes each, except one cluster has only eight cubes to meet construction requirements.

The cubes are corner reflectors, targets for laser-tracking beams. Laser tracking is so precise that data has to be corrected for the 42-inch distance from corner reflectors to the center of the satellite. Extremely brief laser pulses, lasting from 200 picoseconds to one nanosecond, from two to 19 inches in physical length, are transmitted toward the satellite during tracking. Excellent rebounding laser signals are being received and analyzed.

Radio Amateurs Tune In *Fuji*

Thousands of amateur radio operators tuned in during the afternoon of August 12 as *Fuji*, the first ham satellite from Japan, was launched on the same H-1 rocket which ferried *Ajisai* to orbit. An hour after launch, the small communications satellite separated from the H-1. Its powerful radio beacon turned on and starting beeping telemetry toward Earth from 1,000 miles in the sky.

Fuji was made in Japan which has one of the largest and most active ham radio populations in the world. The Japanese traditionally name their satellites after flowers, thus the name *Fuji* which means wisteria. During construction before launch, *Fuji* was known by the utilitarian name of Japan Amateur Satellite One (JAS-1).

Meanwhile, Japan has big plans for the future. It will send the Muses-A satellite around the Moon in 1990. It will add a module for the U.S. space station to be built in space in the late 1990s. It will build its own space shuttle. And it will continue to send scientific satellites to study space.

SOLAR-A Will See Spots

The Sun goes through periods of intense activity followed by quiet periods. These ebb and flow over eleven-year periods in what

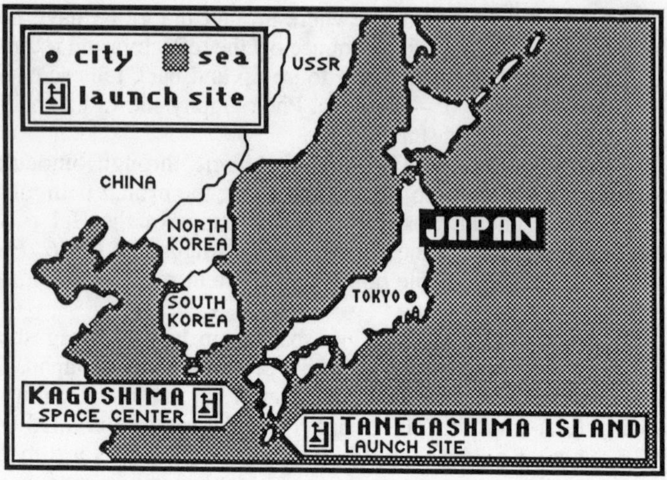

has come to be known as the Sunspot Cycle. During periods of peak activity, the Sun fires off magnificent high-energy solar flares which affect the transmission of radio and television signals on Earth. To study these flares from a better vantage point, Japan will place an eye in the sky called *SOLAR-A*. The craft will be launched into a low Earth orbit in 1991.

SOLAR-A will be one of a series of satellites to be launched by Japan from their Kagoshima Space Center aboard M-3S-II rockets between August and September 1991. That period will be known for highest solar activity during the current Sunspot Cycle. *SOLAR-A*, flying at an altitude of about 344 miles, will stay in orbit for three years. During that time, it will stare straight at the Sun, looking for and examining solar flares on the face of our star.

NASA scientists will supply the optics and detector for a soft x-ray telescope aboard *SOLAR-A*. The Japanese will build the electronics, computer and power system for the satellite. The satellite will see x-rays and gamma rays coming from the Sun. Soft x-rays are weaker x-rays containing less energy.

Japanese Shuttle

Everybody wants a truck to haul stuff out into space and bring it back. The Russians are about to launch a shuttle. The Europeans are planning one. Now the Japanese are studying how to build a shuttle of their own.

Japan's Science and Technology Agency is looking into designs, construction techniques and the question of whether to go it alone or throw in with another country. They could build a small space plane—like a car to carry people and luggage only—to be launched atop the powerful Japanese H-2 rocket expected to be available for use in 1991. A larger shuttle—like a truck to carry cargo—could be built later and launched on an H-2 with extra booster rockets strapped to its sides.

Eventually, the Japanese want a combination airplane and space plane which will take off from, and land on, runways on Earth. That development is expected after the turn of the century.

Satellite Saves Lives First Day

A brand-new *NOAA 10* weather satellite, unlimbering during its first day of operation in space, saved the lives of Rory Johnston and three other Canadians when it found their small plane where it had crashed in a remote area of Ontario.

Johnston's small Cessna airplane lost power during takeoff in poor weather, forcing an emergency landing on a lake. The plane sank, nose down, in eight feet of water.

With face cuts, bruised shoulder and dislocated wrist, Johnston swam 200 yards to shore where he found a canoe. He paddled back out to his sunken craft and brought his injured passengers ashore.

Meanwhile, radio receivers in *NOAA 10,* launched on an *Atlas E* rocket from Vandenberg Air Force Base in California on September 17, 1986, had just been activated during its 76th orbit around the Earth. The same day, during its 90th revolution, special search-and-rescue radio gear on board *NOAA 10* heard Johnston's distress signal.

A satellite equipped with search and rescue equipment, called SARSAT, picks up signals from ships and planes in trouble and relays them to rescue crews on the ground. *NOAA 10's* SARSAT receiver heard Johnston's emergency beacon while flying over Canada. It relayed the distress signal to Canadian rescue forces at Trenton.

At the same time, a Russian satellite equipped with SARSAT also heard Johnston's beacon coming up from Ontario. *NOAA 10* picked up the signal again on its 91st orbit. A pilot of a private plane also reported Johnston's emergency beacon to authorities.

The combination of reports from Soviet and American satellites and the private pilot alerted rescue teams in Edmonton, Alberta. A four-engine C-130 Hercules plane with paramedics aboard took to the air in search of Johnston. Poor weather prevented the C-130 crew from spotting Johnston's plane that night. The next morning, when fog lifted, they found the Cessna and two medical technicians parachuted in to provide first aid.

NOAA 10's report had been so accurate the C-130 crew reported hearing the distress beacon exactly where the satellite said it was. Such airplane distress signals are sent by an emergency locater transmitter (ELT) activated automatically upon impact.

After receiving first aid, Johnston and his three passengers were flown to an airstrip at Sachigo Lake, transferred to another plane and taken to Winnipeg, Manitoba, where they were hospitalized.

The international satellite search and rescue program grew out of efforts in the 1970s by amateur radio operators to listen for emergency beacons through their ham radio satellites in orbit. Today it is an area of international cooperation between the governments of the U.S., the USSR, Canada and France. The U.S. National Oceanic and Atmospheric Administration (NOAA) has two SARSAT satellites in orbit while the USSR has three in operation. NASA says the program, started in 1982, has saved more than 600 lives.

Hurricane Hunter Satellite

After two delays, NASA fired a Delta rocket from Launch Complex 17 at Cape Canaveral Air Force Station, Florida, February 26, 1987, successfully placing the *GOES 7* weather satellite in orbit.

GOES, a Geostationary Operational Environmental Satellite, will be operated by the U.S. National Oceanic and Atmospheric Administration (NOAA).

The *GOES* series of weather satellites have been in stationary orbits some 22,000 miles above Earth. They photograph cloud cover and transmit pictures to weather stations on the ground. The latest *GOES* satellite is eighth in the series. Seven have gone to orbit. One was destroyed at launch in 1986.

The new *GOES* also will report profiles of atmosphere temperatures, collect data on the environment around it in space, and be on the lookout for airplane and ship distress messages.

GOES-7, designated *GOES-H* before launch, was placed into an elliptical or egg-shaped orbit from which it was transferred to a circular orbit on the fourth revolution around the globe. An identical satellite, *GOES-G*, was destroyed May 3, 1986, when its Delta rocket exploded. NASA had a lot riding on the $55 million *GOES-H* as it was the last weather satellite in inventory.

The launch of *GOES-H* was postponed February 24 when a fuel leak was discovered in the Delta rocket. It was postponed again February 25 when high winds threatened rocket flight.

The flight to orbit was not flawless. Two heaters, controlled by thermostats, on the motor which pushed the satellite from low-Earth orbit up to the higher geosynchronous orbit, malfunctioned. The problem was spotted three hours after launch. NASA technicians worked around the problem to put *GOES 7* at its first checkpoint on March 5. The first good test cloud picture from *GOES 7* was received March 6. The infrared picture system was tested successfully March 17. Starting March 18, NASA let it drift east to its proper stationary spot at 75 degrees west longitude over Philadelphia. *GOES* arrived March 24 and drifting was stopped. The satellite now fires small station-keeping jets to maintain its location in the sky.

NASA flight controllers at Goddard Space Flight Center, Greenbelt, Maryland, controlled *GOES-7* for a month while checking out the spacecraft's systems. Control was maintained from facilities in Suitland, Maryland, and Wallops Island, Virginia. After checkout, NOAA Satellite Service flight controllers assumed operations.

First images for public distribution, such as those presented by local television weather forecasters, began March 24.

The infrared system gives *GOES East* its 24-hour-a-day photo capability. It measures temperatures, altitude and energy levels of clouds during nighttime when the primary visual cloud-photo system can't see in the dark.

At its position above 75 degrees west longitude, *GOES-7* is referred to by the weather service as *GOES East*. It provides primary weather coverage for the East Coast of the United States. Its twin, *GOES-6*, known to forecasters as *GOES West*, at 135 degrees west longitude, provides coverage for the western half of the U.S.

GOES-5 had been functioning until 1985 as *GOES East*. *GOES-6* had been working as *GOES West* at 135 degrees west longitude in 1985.

GOES-6 had to be moved in 1985 to the center of the U.S. at 98 degrees west longitude to provide the widest possible coverage of the U.S. from only one satellite after the camera failed on *GOES-5*, the former *GOES East*.

The optical encoder in the imaging and sounding system on *GOES-5* died. Encoder lamps on *GOES-7* have been designed to operate at a lower voltage to increase operating lifetime. A backup encoder with light-emitting diodes also has been installed in *GOES-7*.

Ground controllers started *GOES 6* drifting back toward the West Coast on March 31. It arrived there April 28.

GOES-7 is the tenth meteorological satellite for transmitting cloud cover pictures from stationary orbit to be launched by NASA since 1974. *GOES-7* also makes soundings which are atmosphere temperature profiles. In addition, it collects space environment data and conducts an experiment for detecting from 22,000-mile-high stationary orbit emergency distress signals on the ground.

The search and rescue test uses a 406 MHz radio to determine if instant alert of rescue teams can be achieved. Currently, there are three USSR and two U.S. satellites in near-Earth polar orbits with receivers tuned to the worldwide aircraft emergency beacon frequency of 121.5 MHz. The new plan being tested by *GOES-7* would have the 22,000-mile-high stationary satellite signal the first alert that an emergency was occurring. The low-flying polar satellites then would pinpoint the location of an emergency beacon signal.

Satellite search and rescue (SAR) started in 1983. Since then, it is said 740 lives—mostly Americans—have been saved by rescue teams on the ground alerted by the low-flying SAR satellites. Canada, France, the USSR and the U.S. are involved in satellite search and rescue to save lives of air and sea accident victims.

GOES-7 has solar power-generating cells on its body to feed electricity to its batteries. It is expected to have a useful life of seven years. The main body of the drum-shaped satellite spins. Antennas atop the cylinder point toward Earth. In orbit the satellite is 12 feet long, seven feet in diameter, and weighs 885 lbs.

A new generation of weather satellites is expected to be ready for flight in 1989.

European Weather Satellites

The European Space Agency (ESA) will launch three new weather satellites over four years to replace two old models in orbit.

The *Meteosat* series started with a first satellite placed in orbit in 1977 and another in 1981. The 1981 satellite was expected to work three years but still is going strong more than five years after launch.

European weather satellites are controlled from ESA's Operations Centre at Darmstadt, West Germany. Weather data is fed to national weather bureaus in Belgium, Denmark, West Germany, Finland, France, Greece, Ireland, Italy, Norway, Portugal, Spain, Sweden, Switzerland, The Netherlands, Turkey and Great Britain.

The working *Meteosat* is in stationary orbit 22,245 miles above England. The 705 lb. cylinder is about seven feet in diameter and ten feet tall.

Meteosat is a combined weather observer and radio relay in the sky. Its main payload is a sharp-eyed radiometer which takes infrared, water vapor and visible light pictures of the Earth. In infrared it can resolve objects down to three miles. Visible light photos go down to 1.5 miles. *Meteosat* takes pictures every half hour and transmits them to ESA at Darmstadt for processing. The processed images are sent by radio up to the satellite for relay down to the 16 countries using the weather data.

The new satellites will have the same capabilities plus the ability to relay weather charts and written weather reports to all 16 countries.

Better Weather Monitor

NASA has come up with a satellite thermometer 100 times more sensitive than earlier equipment.

The *infrared radiometer* will measure ocean surface temperatures from space. The new radiometer can tell the difference between the surface of the ocean and the atmosphere immediately above it—a major breakthrough. It was difficult to draw accurate climate maps before because sensors couldn't make the distinction.

NASA put the new gadget aboard the Goodyear blimp *Columbia* and flew it over ocean between Los Angeles and Santa Catalina Island. The radiometer gauges sea temperatures by receiving and analyzing infrared light in sunlight reflected naturally from the ocean surface.

The radiometer is a passive device because it only receives light in the infrared part of the energy spectrum. Radar, on the other, is an active device because it both transmits and receives radio signals to monitor weather patterns.

Meteorologists want the better temperature and wind maps from the radiometer. They will use the data to draw pictures showing the interaction of worldwide weather patterns. For example, they hope to learn how storms start in the tropics and spread to other latitudes. Eventually, they want to understand powerful weather such as *El Nino,* unusual warm water in the eastern Pacific Ocean which can damage fishing industries.

Pacific Satellite Air Calls

Pacstar, in space once before when it was known as *Palapa B1,* will go to space again in 1989 aboard an *Ariane 4* rocket from South America to let airplane passengers over the Pacific Ocean call home.

Indonesia's *Palapa B-1* and Western Union's *Westar 6* communications satellites were ferried to orbit by shuttle *Challenger* in February 1984. They were dropped off in orbits lower than planned so both failed. Shuttle astronauts grabbed them from space in November 1984 and brought them down to Earth. Insurance companies sold them to other satellite operators.

Its new owner refurbished *Palapa,* renaming it *Pacstar.* The satellite will provide communications across the Pacific Ocean including to and from the U.S. West Coast. One service will be telephone calls for airline crews and passengers over the Pacific.

European Space Agency launches *Ariane* rockets from Kourou, French Guiana, South America.

Narrowing The Beam

Big time satellite builders have looked into the future and seen narrow signals beaming to Earth. Communications satellites are about to sprout lots more antennas.

Once upon a time telephone calls went from point to point via a network of wires. Microwave radio signals later carried additional calls. Then communications satellites brought a big leap in the number of calls which could be handled at one time. Now we have telephone transmission cables woven from hair-like strands of glass called fiber optics.

It seems AT&T and other down-to-Earth communicators have pushed fiber optics to the point where the new technology offers greater capacity than satellites. Some say satellite communications business might wind down as fiber optics networks are built up.

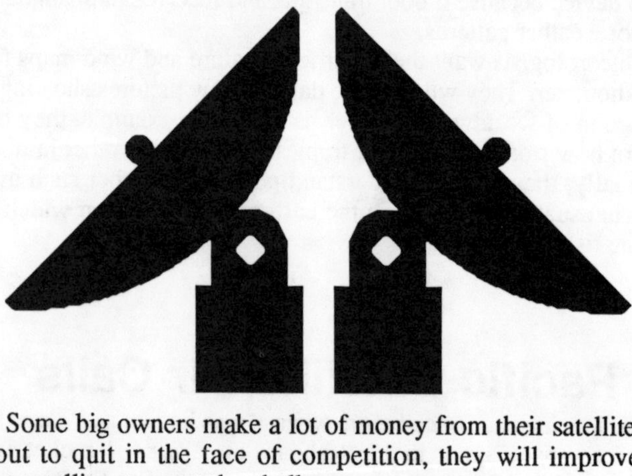

Some big owners make a lot of money from their satellites. Not about to quit in the face of competition, they will improve their next satellites to meet the challenge.

Until now each satellite has used one antenna to send down a beam broad enough to cover the entire USA. The new idea is to build a bird with several antennas, each beaming down to a part of the country. Instead of one giant radio footprint covering the whole country, many overlapping footprints would blanket the states.

Tightly focusing a radio beam increases the strength of its signal arriving on Earth. Smaller receiving antennas would be required so broadcasters, governments, corporations and other users would find satellite communications cheaper. Where it had been necessary to build a dish antenna of 45 to 100 ft. diameter, a 5 to 10 ft. antenna would suffice, saving money. Eventually, it could become possible to send different information down each beam, increasing the capacity of the satellite.

Forest Fire Satellite

Firefighters battling blazes 20 miles inside a western forest in August 1986 used a sailor's satellite to call out for backup and supplies.

The International Maritime Satellite Organization's *Inmarsat* satellite was pressed into service as 18,000 fighters tried to douse raging fires which burned 700,000 acres of wilderness in Oregon and Idaho.

Commanders in the field typed messages on small lap-top portable computers. The computers were hooked to portable two-way radios. Using cumbersome dish antennas, the radios transmitted messages requesting shift changes and supplies up to the satellite. It retransmitted the signals down to receiving stations at firefighter headquarters several miles away.

Workers usually use handheld radios and telephones to report into basecamps but some of the fires this year took fighters so far into the woods no telephones or battery rechargers were nearby.

It's not cheap to use modern technology during forest fires. The portable satellite ground stations cost $40,000 to $60,000 each. In addition, firefighters have to rent the satellite during the time they use it. Also, it took almost two days to set up the satellite system in the field.

Ship-To-Shore Satellite

Some 3300 ships at sea—tug boats, passenger liners, fishing boats, oil tankers, liquid natural gas vessels, scientific research ships and even offshore drilling rigs—talk to folks back home and other ships at sea using the International Maritime Satellite Organization (INMARSAT) spacecraft in Earth orbit.

The world's first maritime satellite for private shipping and offshore industries was built in 1976.

Ten years later, a half dozen improved satellites are set for launch over the next few years. Each of the new satellites will carry 125 simultaneous radio transmissions.

South Pole Satellite

U.S. scientists exploring the South Pole now can call home directly through an 18-year-old satellite passing overhead.

Engineers who built the third Applications Technology Satellite *ATS 3* thought it would work seven years in orbit. To their pleasant surprise it kept on working and has been in stationary orbit for 18 years. But now it has run out of fuel to keep it on location so it drifts around the globe.

Every day it drifts to a spot where it can be seen from South Florida and the South Pole at the same time for about four hours. A ground station at the University of Miami and a new NASA ground station near the South Pole keep scientists talking back home.

Plane And Train Phone Calls

If the Federal government approves, you may be able to use a space satellite to make a telephone call directly to a plane in flight or from a moving railroad car.

Your long-distance call would be radioed directly from your train, plane or automobile up to a satellite in Earth orbit—even while you are in motion. The satellite would relay your real-time telephone call down to make the connection to your home or office or another plane, train or car.

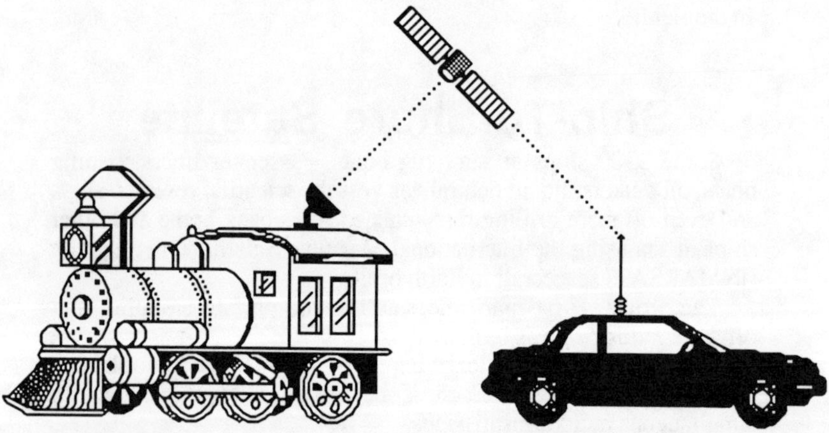

Computer data could be sent through the mobile system as well as ordinary voice calls.

Portable gear would permit telephone calls to and from remote locations such as mountain tops, deep forests or mid-ocean vacation retreats. The satellite could cover the U.S., Canada and Mexico.

Maritime Satellite

With 300 large ships plying the oceans, East Germany has become the 47th country to sign on with *Inmarsat*, the International Maritime Satellite Organization.

Now the East Germans can outfit their freighters with dish antennas on decks and satellite receivers in radio rooms. Wherever they sail, they will be able to send and receive electronic mail, telephone calls and navigation positions.

Inmarsat, based in London, operates three satellites in stationary orbits. Even pleasure craft and small fishing boats can use the system.

Some 5,000 vessels now have large dish antennas and $13,000 radio setups to transmit and receive messages through the satellites. *Inmarsat* is expecting the number of boats using the satellites to increase now that it has an inexpensive radio the size of a CB set requiring only a small whip antenna.

At $5,000, even sailboat owners are expected to add the new whip antenna to their masts and the tiny 13-lb. radio to their cabins. European manufacturers started turning out the new electronics in 1987.

Marine radios sometimes have been less than reliable. High-frequency ship-to-shore radios are affected by weather. Ship-to-ship VHF is limited to line of sight.

Satellites, on the other hand, are so high in the sky they can see down on an entire hemisphere. A ship in the middle of an ocean and a station in an office on shore both can see the satellite at the same time. All that's needed is a small VHF radio to sends signals up to the satellite. The satellite retransmits the signal down to Earth. Anybody who can see the satellite can receive the down-link transmission. VHF signals sent to satellites are not affected by weather.

Bigger Phone Satellite Coming

The International Telecommunications Satellite Organization (Intelsat) is readying five giant communications satellites, each capable of carrying 120,000 telephone calls and three television broadcasts at the same time.

The series of *Intelsat VI* satellites will be the biggest and most powerful in orbit. Their spot radio beams will give better coverage of North America and better connections to Europe.

An *Intelsat VI* is 39 feet tall. Its radio signal is so powerful, a receiver on the ground requires a dish antenna only two feet in diameter.

That compares with *Intelsat I* of 20 years ago which was slightly more than four feet tall. Back then, receivers on the ground needed dish antennas 100 feet in diameter.

Swedish Mailstar

Sweden is said to have more international companies per capita than any other country. Those companies need to send messages to branches on all continents.

To serve those international firms, the government of Sweden and three private companies there are building a pair of small inexpensive lightweight communications satellites for launch atop a Chinese *Long March* rocket in 1989.

The *Mailstar* satellites will use modern packet-radio technology, much like the *Pacsat* craft reported elsewhere in this book.

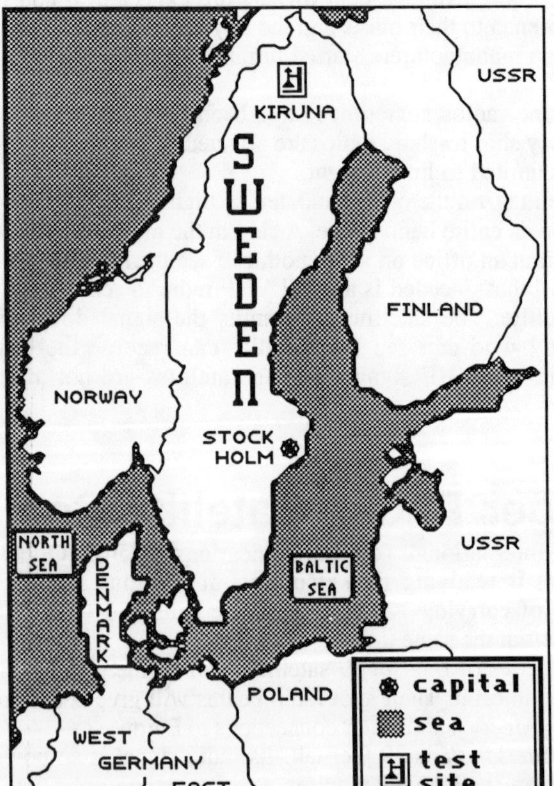

Mailstar craft will be about 500 to 900 miles out in space in an orbit which will take them over the Earth's poles fourteen times a day. Each will pass over any one point on the globe for about ten minutes from six to fourteen times a day.

Typewritten messages stored in small personal-size computers at ground stations will be sent up by radio to a satellite when it is overhead. Later, other ground stations can command the satellite by radio to send down stored messages. Each satellite will be able to store 80 million characters in messages at one time. Ground stations are expected to cost under $10,000 with many in use around the globe.

Three companies working with the Swedish Telecommunications Administration on the project are Saab Space, Ericsson Radio Systems and the Swedish Space Corporation. They have reserved *Long March* flights in 1989, 1992, 1995, 1997, 2000, 2002 and 2005.

The satellites will be controlled from Kiruna, in northern Sweden, where the Swedish Space Corporation has a rocket firing range. Each satellite will be in view overhead of the Kiruna ground controllers for 10 of 14 trips around the world each 24 hours.

Space Age Documents

Efficient communications between countries is important to their economic, political, social and cultural development. When the post office is too slow, ten European nations now will switch to sending electronic documents through space satellites.

All kinds of government documents will be easily and quickly accessible throughout Western Europe. Belgium, Denmark, Ireland, Italy, The Netherlands, Norway, West Germany, Spain, Sweden and Great Britain are supporting the new *Apollo* program to send documents through space for speedier delivery. The British Library document supply centre at Boston Spa in West Yorkshire, England, in January 1987 became the first to send electronic paperwork through the *Eutelsat F2* satellite.

Mexican Satellites: Morales

Mexico named its communications satellite system after Jose Maria Morelos y Pavon, a hero from their revolution which brought independence from Spain. He wrote a declaration of independence from Spain in 1813 calling for racial equality and government by the people.

Palapa Launched On Delta

NASA made it two in a row in less than a month with the successful launch March 20, 1987, of Indonesia's *Palapa B2-P* to orbit from Cape Canaveral.

Palapa, second of three spacecraft in the *Palapa B* series, rode a *Delta* rocket to space. It will add 24 telephone, television, telegraph and data communications channels for Indonesia and the Association of South East Asian Nations (ASEAN) countries. ASEAN includes the Philippines, Thailand, Malaysia, Singapore and Papua, New Guinea.

Palapa B1 was ferried to orbit by a shuttle in June 1983. *B2* was carried to orbit, along with Western Union's *Westar VI* satellite, by shuttle *Challenger* in February 1984. They were dropped off in orbits lower than planned so both satellites failed. They were retrieved by shuttle in November 1984 and brought down to Earth.

The name *Palapa* signifies unity, commemorating the unification of the Indonesian nation by advanced satellite communications.

GOES 7, the U.S. East Coast weather satellite, was sent to orbit from Cape Canaveral February 26, 1987, aboard a Delta rocket.

Aussat and *Brazilsat*

Aussat 2, the second of three Australian communications satellites, is in stationary orbit 22,300 miles above that Southern Hemisphere continent.

Three antennas beam radio, television, telephone and data signals down to the continent and offshore islands. Even Australian air traffic controllers use the satellites.

Meanwhile, *Brazilsat 2,* a television, telephone and data communications satellite to link the Amazon jungles with cities in Brazil, has joined *Brazilsat 1* in orbit over the South American continent.

USSR TV Satellites

Live Russian radio and television programs can be received in North America via satellite.

Four USSR *Molniya* television satellites chase each other around the globe about six hours apart in high elliptical orbits. *Molniya* is Russian for *lightning.*

U.S. tv satellites are in stationary orbits, each beaming down from its same spot overhead at all times. The USSR satellites, on the other hand, are not stationary since such orbits would limit their reach to about 70 degrees north latitude. Moscow wants to reach provinces farther north so satellites moving in elliptical orbits are used. The four satellites provide 24-hour coverage to the far north as well as the rest of the vast USSR land mass.

The orbital path permits reception in the U.S. of 15 hours of television and 24 hours of radio programs each day.

One North American audience can be found in the United States Information Agency (USIA) offices in downtown Washington, DC. USIA has a computer-controlled tracking antenna on its roof, looking for whichever *Molniya* is in view. Based on known orbits, the computer points the antenna toward the spot in the sky where a *Molniya* satellite is likely to be found.

From Moscow With...

The USSR is about to use a satellite to relay Moscow television programs into Poland.

Agreeing to accept Russian re-runs is a price the Polish government paid for USSR backing at a Communist Party congress in 1986. The price probably isn't too great since Moscow programs usually go over like a lead balloon in the ratings.

Voice Of America Via Space

The United States Information Agency (USIA) will be replacing two Nazi transmitters captured at the end of World War II with space satellites to send the Voice Of America (VOA) around the world.

VOA broadcasts news and entertainment programs in 42 languages to local audiences all over the world from transmitters in a chain of countries around the world. Programs are sent from Washington, DC, via 12 high-frequency point-to-point earthbound radio relay stations. Each local VOA station around the world broadcasts many different programs at the same time on different frequencies in different languages.

By the early 1990s, USIA wants to improve VOA's sound quality and save money at the same time by relaying the programs to local transmitter sites around the globe via satellite. COMSAT, the private communications satellite corporation, will design and install the VOA network. It will include two-way radio via satellite so VOA in Washington can tell if one of its transmitters at a remote location has gone off the air.

The former Nazi equipment now is in use in Germany.

Japan's Line To America

Japan is propagandizing the eastern United States by satellite.

Station NHK, known as *Radio Japan*, transmits the government line around the world—usually via shortwave radio—much the same as the *Voice of America* sends U.S. propaganda to foreign countries.

The first problem: a weak shortwave signal from Japan turned off listeners in the eastern United States.

The first solution: Japan enlisted the help of Canada in targeting listeners who weren't tuning to NHK. *Radio Canada International*, Canada's propaganda arm, would pick up NHK shortwave transmissions and rebroadcast them in a beam aimed at the eastern United States.

The next problem: *Radio Canada International* also had trouble copying NHK on shortwave.

The next solution: Rather than put its signal for U.S. East Coast listeners out on shortwave, NHK would convert them to microwave signals and beam them straight up to an *Intelsat* satellite in stationary orbit over the Pacific Ocean. *Intelsat* would retransmit them down to a satellite receiving station at Vancouver, British Columbia. From Vancouver they would be sent through telephone lines and microwave all the way across Canada to New Brunswick. A giant 250,000 watt *Radio Canada International* shortwave transmitter in New Brunswick would beam NHK's programs south along the U.S. East Coast. Listeners tuning their shortwave radios to 6.120 MHz hear NHK between 6:30 a.m. and 7:30 a.m. eastern time.

The question: how many think the broadcast is coming all the way around the world from Japan rather than only a few hundred miles from Canada?

Prospecting By Satellite

The Boy Scouts of America took the high road in search of treasure in 1986.

Fearing tax changes would cut down on donations in 1987, the scouts decided to hit up big spenders in 400 BSA councils across America before the end of 1986. Rather than waste time going door to door, they sent a signal up to a communications satellite in orbit.

The message was beamed down to 3,000 potential givers clustered in 86 Holiday Inn motels around the country. One Memphis cotton merchant who saw the show was so impressed he kicked in $500,000.

Ten Four, Good Buddy

Truckers herding 18-wheelers across the byways of America now can keep in touch via satellite link from cab to home office.

A Los Angeles company socks truckers $3,500 to put a tennis-ball size antenna on the cab and a video screen beneath the dash. Dispatchers will know exactly where each truck is. They can redirect drivers to stop off for new pickups as loads become available along their routes.

For $2 a message, big supermarket chains can send drivers after fresh cabbage and express companies can expedite their parcels. The big question: will truckers talk plain English through satellites or resort to Big Bopper handles and crude codes to get messages through?

Financial Satellite

A New York financial company has its computer in Alaska while its offices are in Maine, Oregon and New York. It keeps in touch with all points by satellite.

The Albany firm uses fifteen-foot-wide dish antennas at its data processing centers in Anchorage and Fairbanks to signal its offices in Augusta, Maine, Portland, Oregon, and Albany, New York via satellite. Data coded into radio transmissions is sent up to a satellite in Earth orbit. The satellite repeats them down to the offices on the ground.

Coast-To-Coast Beeper

A new National Satellite Paging (NSP) system makes it possible for a beeper in the pocket of a salesman driving on a Los Angeles freeway to receive an alert from his home office all the way across country in New York.

Like other pagers with tones urging you to call home, NSP sends out a radio signal which makes your receiver beep. The difference is the message can originate in New York, L.A., Chicago, Washington, San Diego, Atlanta, San Jose, Houston, Boston, San Francisco, Austin, Miami, Dallas, Denver, San Antonio or Jackson Mississippi, and travels to you via satellite.

When NSP receives a message in one city, it sends a signal up to a satellite in earth orbit. The satellite relays the signal down to the destination city on Earth. A local transmitter in the destination city is turned on and a pager alert is sent out. The beeper sounds off if it is in that transmitter's coverage area.

NSP service is planned for 100 other cities. It gives business travelers a nationwide network of inner city coverage. Most beeper services cover only local territories.

Ashes To Space...

It used to be ashes to ashes, dust to dust. Now a Florida funeral home wants to send cremated remains to space.

A private rocket company in Houston may launch a mausoleum in 1988. It will house 10,000 tubes of remains. The funeral home claims 3,500 people have signed on at $4,000 per tube.

Old-time Mercury astronaut Donald K. "Deke" Slayton heads Space Services Inc., the private launch firm. He sent a private rocket on a suborbital flight in 1982.

Space Services has rockets on hand to launch small satellites up to 3,000 pounds from Virginia, Mississippi or Hawaii. Communications or environmental data-gathering satellites could come in under that weight. The funeral home is Celestis Group Inc. of Melbourne, Fla.

Polar BEAR Launched

The Air Force's *Polar BEAR* satellite made it to orbit in November 1986 aboard a NASA *Scout* rocket after two delays.

The $13 million communications satellite will photograph the Northern Lights while studying electrical particles and the Earth's magnetic field over the planet's poles.

Polar BEAR stands for Polar Beacon Experiment and Auroral Research. It's a 20-year-old Navy radio satellite which had been mothballed at the Smithsonian Institution. When the Navy built it as a spare for backup, it was known as *Oscar 17*. The 270 lb. satellite was not used so it was placed on display in the Air and Space Museum 15 years ago. The Air Force traded an extra *Transit 5A* satellite for the old *Oscar 17*. Technicians sprucing up the old bird found electronic equipment in good working order. Only one solar cell out of thousands on the satellite's solar power panels had to be replaced.

Polar BEAR was launched into a 625-mile-high polar orbit from California aboard a 75-foot-tall *Scout* rocket.

All satellites depend heavily on radio to receive commands from Earth and to send data down from space. But the Northern Lights aurora borealis create static which interferes with communication when satellites are near the North Pole. Polar BEAR will take a close look at the problem.

The former Navy *Oscar* program is not the same as the 25-year-old ham radio Orbital Satellite Carrying Amateur Radio—*OSCAR*.

Salvaged Satellite On The Job

Remember the time shuttle astronauts jury-rigged a flyswatter in an attempt to turn on a communications satellite which failed to start up in orbit?

Well, the satellite is on station at a spot 22,300 miles out in space and working fine.

The Leasat F3 satellite was pushed out of the shuttle in April 1985 but it failed to start. Astronauts couldn't get it going so, four months later, other astronauts gave it a jump start. That allowed ground operators to regain control. It was parked in low-Earth orbit for two months and then a motor was fired which drove it out to 22,300 miles. It was tested at that geostationary orbit and turned over to the U.S. Navy.

Satellite Finds Ice Station

A Russian radar satellite known discreetly as *Cosmos 1766* found the drifting and lost Druzhnaya Antarctic ice station in 1986.

The station crew had set up shop on ice at Filchner Glacier. They were lost when the ice broke off and drifted away. A side-looking radar on the oceanographic satellite spotted the drifting ice floe and revealed a sea path to save the crew.

Satellite Map In Your Car

A picture tube in the automobile dash displays a map of city streets surrounded by country roads. As the car drives along a street, a large black dot hops along the map. A trail of dashed lines tell where the vehicle has been. A radio receiver behind the dashboard hears and understands signals from a *Navstar* Global Positioning Satellite some 12,532 miles above in Earth orbit.

Navstar will be a constellation of 21 satellites, including three spares standing by for emergencies in orbit. So far, seven have been sent to space. Each satellite circles the globe twice a day. Four satellites are in view from any spot on Earth at any time.

With four satellites looking down on a receiver, three-dimensional navigation is possible. A receiver tuned to the satellites can determine its own longitude, latitude and altitude. The system can be used by planes and ships as well as land vehicles.

The heart of *Navstar* is its accurate clock. Each satellite has an atomic oscillator with rubidium cell and cesium beam. Clock accuracy is within one second in 100,000 years! Navstar Control Center at Colorado Springs, CO, is part of the U.S. Air Force's space command. It checks the satellite clocks regularly to make sure they are beating in sync.

The satellites beep away cheerily, 24 hours a day, with never-ending messages telling what time it is. A receiver in a car checks its own clock, comparing it with time signals from four satellites to compute how far away the satellites are.

Manufacturers already are demonstrating automobile navigation systems based on Navstar. Price of car receivers should drop below $500 by 1992. They may combine a safe heads-up display projected on the windshield with a computer which understands spoken words. You would say aloud street names. The gear would respond by showing how to get there.

In fact, by 1992 receivers may be so small that one-pound handheld personal navigators may be available for backpackers to lug into the wilderness. Receivers in small planes will permit in-the-dark landings within ten feet of intended touchdown. Police and fire vehicles could be tracked continuously by dispatchers. Nationwide tracking of refrigerator railcars and uranium shipments would be possible. Truck, train and ship terminal bottlenecks might be eliminated because dispatchers would know ahead of time who is coming in when.

Ships at sea and on inland waterways could avoid collisions with a warning device built around Navstar. Crop dusting aircraft could be run entirely by autopilot using location information from Navstar. Similarly, road grading could be done by a robot equipment operator receiving Navstar location information. Downed aircraft or boats lost at sea could be located readily if they transmitted positions based on Navstar receiver calculations. Robots could do land surveys. Ocean buoys could become self-monitoring. Air traffic control might become automatic. Map making and wild-animal tracking could be completed instantly by machine.

Engineers are predicting all ships, airplanes and most automobiles will include Navstar receivers by 1999.

Spy Satellites

Peeping Tom photos and radio eavesdropping are the latest rage in spy satellites.

The best U.S. spy eye in the sky is cryptically referred to as KH-11. Like a tv camera looking down from space, it converts pictures into digital radio signals beamed to Earth.

Ears in the sky are called ELINT for ELectronic INTelligence satellites. They are radio receivers in the sky to tune in on telephone conversations on the ground, radar signals, missile telemetry signals, and other messages.

An improved KH-11 called KH-12 will be sent up to its duty station when U.S. shuttles are ready again to carry freight to space.

Amateur Radio Satellites

Two new Russian ham radio satellites were delayed at launch during the winter of 1987 when cold weather gripped the Northern Cosmodrome at Plesetsk.

Plans call for the two radiosputniks—called *RS 9* and *RS 10*—to be ferried to orbit in 1987, piggybacking on one rocket when the USSR launches a navigation satellite.

The Russian amateur radio sputnik *RS-10* was expected to be launched in June 1987. *RS-9*, on the other hand, has been postponed with no future launch date announced. Both *RS* satellites first were expected to go to space at the start of 1987.

Meanwhile batteries in two older Russian amateur satellites still transmitting from space, *RS 5* and *RS 7*, are growing weaker but the radio systems still work.

Amateur radio satellites have been built and sent to orbit several times in the past 26 years by private groups of radio operators. Those satellites were built by hams and flown as free-ride ballast on various government rockets carrying other commercial and government satellites to orbit. Many remain in orbit today (see the table below).

USA radio amateurs had international help from amateurs in other countries—West Germany, Japan, Canada, Australia, Great Britain, and elsewhere—during design and construction of the *OSCAR* series. *OSCAR* stands for Orbital Satellite Carrying Amateur Radio.

The U.S. and Western Europe's *AMSAT Phase 3C*, to be known in orbit as *OSCAR*, may be launched in early 1988 by the European Space Agency on an *Ariane 4* rocket from Kourou, French Guiana.

AMSAT is the Radio Amateur Satellite Corp., a worldwide club of amateur radio space buffs.

USSR students built *Iskra 2* and *Iskra 3* which were launched by hand from the USSR *Salyut 7* space station in orbit. *Iskra,* in Russian, means *spark. RS 9* and *RS 10* are the two expected to be launched in 1987. *RS* stands for *Radio Sputnik* or radio satellite. *RS 1* also has been called *Radio 1* and *PC 1* due to confusion in translating from the Russian.

They should be launched from northern Russia into circular orbits about 650 miles above Earth.

If you have a shortwave receiver, try listening for the radio beacons coming from RS 9 and RS 10 after they are launched. You might hear RS 9 transmitting International Morse Code on a frequency of 29.402 MHz. RS 10 beacons may be at 29.457, 29.503, 145.957, or 145.997 MHz.

These satellites are used by hams on the ground to talk with other amateurs on the ground. The radio equipment in the satellites

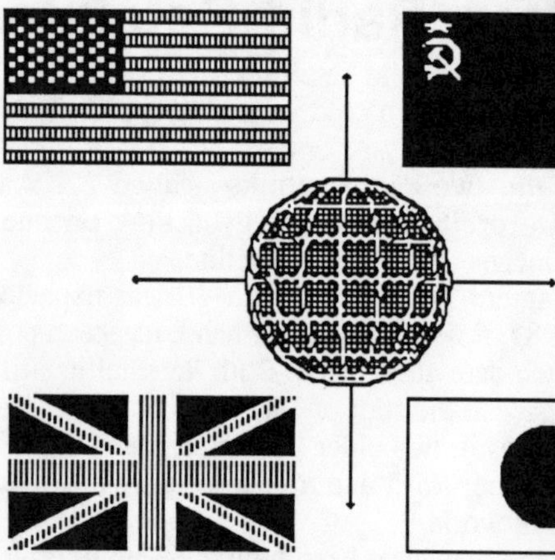

relay signals over great distances. Such relay signals coming down from RS 9 will be in the frequency range from 29.360 to 29.400 MHz. RS 10 transmits downlink signals from 29.460 to 29.500 and 145.960 to 145.995 MHz.

In addition, the satellites have computers which act as robot radio stations. A satellite's receiver listens for hams on the ground sending up signals addressed to the robot. When such a signal is recognized, the robot replies with a greeting and a report on the strength of the signal coming up from the ground. When the satellites are in orbit, these robots may be heard on 29.320, 29.457 or 29.503 MHz.

NUSAT 1, strictly speaking not an amateur radio satellite but an amateur satellite, was built by students at Weber State College in Utah and launched from *Challenger* while the shuttle was in orbit. *NUSAT* stands for Northern Utah Satellite. *NUSAT 2* is being readied for launch in 1988 and may carry an amateur radio station.

UoSAT stands for University of Surrey Satellite. The two in orbit were built by students at that university in England.

OSCAR 5, also known as *Australis-OSCAR 5*, was designed and built by students at the University of Melbourne, Australia, and flew aboard a U.S. rocket.

Ham operators in France and other countries are planning space satellites of their own. The French spacecraft, when and if launched, may be called *ARSENE* and could be in an elliptical orbit like *OSCAR 10*, which went to orbit on a French rocket.

Most of the amateur radio satellites listed below are in low orbits near Earth. They circle the globe, coming within range of any one ham station on the ground every hour or so. *OSCAR 10*, on the other hand, is in a long elliptical orbit which keeps it in view of many ground stations for hours at a time. Most ham satellites have transponders which act as radio signal relays in the sky. A ham operator on the ground transmits his signal up to the satellite which repeats it downward over a wide area.

Amateurs are planning geosynchronous stationary satellites for the future. And a ham radio station is being planned for the U.S. space station when it is launched in the late 1990s.

Here's a list of amateur radio satellites over the years. All remain in space except for *OSCAR 1*, *OSCAR 2*, *OSCAR 4*, *Iskra 2*, *Iskra 3* and *NUSAT 1*:

SAT	LAUNCH	COUNTRY	COMMENTS
OSCAR 1	Dec. 12, 1961	USA	operated 21 days
OSCAR 2	June 2, 1962	USA	operated 19 days
OSCAR 3	Mar. 9, 1965	USA	lasted months
OSCAR 4	Dec. 21, 1965	USA	operated 85 days
OSCAR 5	Jan. 23, 1970	Australia	operated 52 days
OSCAR 6	Oct. 15, 1972	USA	operated 4.5 years
OSCAR 7	Nov. 15, 1974	USA	operated 6.5 years
OSCAR 8	Mar. 5, 1978	USA	operated 5.3 years
RS 1	Oct. 26, 1978	USSR	also called Radio 1
RS 2	Oct. 26, 1978	USSR	lasted months
OSCAR 9	Oct. 6, 1981	UK	UoSAT still operating
RS 3	Dec. 17, 1981	USSR	operated years
RS 4	Dec. 17, 1981	USSR	operated years
RS 5	Dec. 17, 1981	USSR	still operating
RS 6	Dec. 17, 1981	USSR	operated years
RS 7	Dec. 17, 1981	USSR	still operating
RS 8	Dec. 17, 1981	USSR	operated years
Iskra 2	May 17, 1982	USSR	operated 53 days
Iskra 3	Nov. 18, 1982	USSR	operated 37 days
OSCAR 10	June 16, 1983	USA	still operating
OSCAR 11	Mar. 2, 1984	UK	UoSAT still operating
NUSAT 1	Apr. 29, 1985	USA	Northern Utah Satellite
OSCAR 12	Aug. 12, 1986	Japan	still operating
FORTHCOMING LAUNCHES			
RS 9	1987	USSR	Radio Sputnik
RS 10	1987	USSR	Radio Sputnik
AMSAT 3C	1987	USA	to be called OSCAR
NUSAT 2	1988	USA	Northern Utah Satellite
JAS 2	1988	Japan	to be called OSCAR

Radio Amateurs Tune In *Fuji*

Thousands of amateur radio operators and shortwave listeners (SWLs) tuned in during an August afternoon in 1986 as *Fuji*, the first ham satellite from Japan, was launched on the H-1 rocket.

An hour after launch, the small communications satellite separated from the H-1. Its powerful radio beacon turned on and starting beeping telemetry toward Earth from 1,000 miles in the sky. A team of hams from AMSAT, the Radio Amateur Satellite Corp., first heard *Fuji's* signals at Santiago, Chile. Half an hour later European amateurs were able to tune in to *Fuji,* followed by U.S. hams an hour later.

The H-1 vehicle was launched from Japan's ultra-high-tech Tanegashima Space Center on an island off the coast of South Kyushu, Japan. AMSAT set up a worldwide radio network to broadcast the launch and first beacon reception to amateurs and SWLs.

AMSAT is a worldwide organization of amateur radio operators, headquartered near Washington, D.C., interested in building and using space satellites for communications. *Fuji* was made in Japan by JAMSAT, the local AMSAT group, affiliated with the Japan Amateur Radio League (JARL). Japan has one of the largest and most active ham radio populations in the world.

The Japanese traditionally name their satellites after flowers, thus the name *Fuji* which means wisteria. During construction, before launch, *Fuji* was known by the utilitarian name of Japan Amateur Satellite One (JAS-1).

Beacon can be heard

Fuji's radio beacon can be heard, whenever it's overhead, sending telemetry in Morse code at 20 words per minute on a frequency of 435.975 MHz. Aboard the satellite are two communications systems. One receives and retransmits voice and Morse code amateur radio signals permitting hams to chat over wide areas of the globe.

The other is a mailbox in the sky. It receives typewritten messages sent up from individual ham stations, stores those messages, and delivers them on request to other stations at later times. This computerized store-and-forward packet radio system permits amateurs on one side of the world to place messages on an electronic bulletin board in the satellite to be removed and read by others when the satellite is on the far side of the world.

Fuji is open for use by amateurs anywhere in the world. Its communications systems receive signals in the 146 MHz amateur band and retransmits them in the 435 MHz ham band. Amateurs have been chatting away happily through *Fuji's* systems since right after launch.

More than a dozen amateur radio communications satellites have been placed in orbit during the last 25 years. Known as *OSCAR*, for Orbital Satellite Carrying Amateur Radio, they have come from the U.S., Great Britain, USSR and Japan, with help from hams in several countries including West Germany, Canada, Australia, France and others. Officially, *Fuji* is OSCAR 12. However, the

number sequence is out of order since a half dozen Russian ham satellites in the RS series were not given OSCAR numbers. RS stands for *Radio Sputnick.*

USSR amateur satellites

Amateur radio satellites are popular among hams in the USSR. Six were orbited by the USSR in 1981. Two, *RS-5* and *RS-7*, still work and are used regularly, although batteries are weakening. Amateur radio satellites from all countries have been expected to function properly for about three years when launched. The two Russian satellites still in use have gone beyond their expected lifespan. Replacement satellites are *RS 9* and *RS 10.*

How ham satellites work

From the point of view of a person standing on the ground, a satellite either is always overhead in the same spot in the sky or it passes by regularly overhead.

The tv broadcast satellite, which has led to a nationwide sprouting of large backyard dish antennas, appears to be stationary in the sky. It is about 22,000 miles above the surface of Earth and moving around the globe at the same speed as Earth is turning. It always seems to be over the same spot on the ground. Dish antennas on the ground remain pointed in one direction to receive signals from the satellite. Television is nothing more than a radio signal with video as well as audio so receivers and transmitters in tv broadcast satellites are pretty much the same as amateur satellites.

Unlike stationary satellites, amateur radio satellites don't travel as fast as the Earth turns so they appear to pass overhead. Ham satellites usually are 500 to 1,000 miles above the surface of the Earth. That's the same or slightly higher than a typical space shuttle flight orbit. To a viewer on the ground, an amateur satellite seems to come around the Earth every hour or so. Sometimes a satellite will be west of the viewer, sometimes directly overhead, sometimes east. One skill learned by an amateur radio operator is how to know where to point his antenna for best reception of signals from a satellite. AMSAT publishes information on orbital paths of each amateur satellite. A ham can use that data with pencil and paper or a small pocket calculator to determine the exact location of a satellite at any moment.

To use a satellite, an amateur operator determines its location and, if it's within sight of his station, points his antenna toward it. He receives signals from many other hams relayed to Earth by the satellite. When he hears an amateur, using the satellite, with whom he would like to chat, he transmits his voice or code up to the satellite which retransmits it down to all other stations.

Radio signals used for such satellite communications usually are very high frequency (vhf) or ultra high frequency (uhf). That means they are good only for line-of-sight paths so communications between any two stations is possible only as long as the satellite is in view of both stations.

USSR, British, Japanese and most U.S. satellites have been of this lower-orbit kind. However, one U.S. satellite, *OSCAR 10*, flies out about 20,000 miles in an odd orbit which brings it close to Earth and then far out again. Such an elliptical orbit makes the satellite seem to stay over some ground stations for many hours before moving out of view. Low-orbit satellites are in view of any one station less than half an hour.

Space ballast

Cost of designing and building amateur radio satellites is borne by hams. The satellites usually get a free ride into space as ballast or dead weight used to balance the satellite compartment on a rocket. Most have been launched by NASA in the U.S. but some have been launched by the Russians and by the European Space Agency at its space center in South America. Japan launched *Fuji* on its own H-1 rocket.

Along with two-way communications, all amateur satellites carry one-way beacons which send down information about a satellite to anybody interested in receiving the data. *Fuji*, for instance, constantly monitors its own solar cells and battery. It transmits information about the amount of current being generated by the solar cells, the voltage available from the battery, the temperature of the battery, transmitter power, and temperatures of other parts of the satellite. This telemetry data is easily read by amateurs.

25th anniversary

A ten pound satellite, called *OSCAR 1*, was launched December 12, 1961. As it beeped down a tracking signal it launched amateur radio into space. Over the past 25 years, thousands of ham operators have used the many successors to *OSCAR 1*. AMSAT members around the world celebrated on December 12, 1986. Information on amateur satellites is available from AMSAT at Postoffice Box 27, Washington, D.C. 20044.

Stationary Ham Satellite

Until now, the dozen and a half amateur radio *OSCAR* satellites orbiting the Earth have not been geosynchronous. They sail around the globe on regular paths which bring them over ham stations on the ground at regular intervals.

Now the amateur radio community is planning an exciting new "bird" which will always be in view of most of the northern hemisphere. Known by its working name of *Phase 4*, the new satellite would be ferried out to 22,000 miles from earth where it would stare down from a stationary orbit, much like the well-known television relays in orbit.

One of the most important functions of the geosynchronous stationary satellite will be to relay communications between hams on Earth and the U.S. Space Station. Amateurs also will talk to each other from point-to-point across the globe through the satellite. Voice, Morse code, television and computer data will sent up from Earthbound ham shacks to the satellite for retransmission toward Earth. A ham operator on one side of the Earth should have a highly reliable path for his radio signal to remain in constant contact with an amateur on the other side of the world. An emergency alerting system to call up amateurs in natural disasters probably will be built into the satellite.

Simplicity and ease of use are planned features of the Phase 4 satellite. Amateurs on Earth will use small dish-shaped antennas about three feet in diameter to transmit signals to, and receive from, the satellite. Once the ham points his antenna toward *Phase 4*, it will not have to be moved no matter where in the world the amateur is communicating.

OSCAR stands for Orbital Satellite Carrying Amateur Radio. Amateur satellites have been built with volunteer effort and money contributed by hams. They have been launched as ballast on U.S. and European rockets.

Packet-Sized Earth Station

A computer, transmitter and antenna—everything needed for an Earth station—now can fit in a suitcase. Soon scores of battery-portable communicators will link underdeveloped countries to weather, agriculture and engineering databanks around the world.

PACSAT, the nickname for a new mailbox-in-orbit packet-radio satellite being built by the U.S. Volunteers In Technical Assistance (VITA) and the worldwide-membership Radio Amateur Satellite Corporation (AMSAT), will make the sending of information in and out of developing regions cheap and easy.

Volunteers helping out in depressed African, Asian and South American areas will use miniature battery-powered Earth stations to ask for and receive all-important technical assistance. Travelers will be able to send and receive data from the most rugged terrain. Relief workers will be able to communicate directly with emergency teams at natural disaster sites.

VITA has been sending technical information through the mails to developing countries for two decades. Now the material can be more timely and reliable as it is sent electronically to the orbiting mailbox where recipients around the world can retrieve it at their convenience.

Computers talking to computers

Packet radio is a two-way communications system where radio replaces telephone lines and computers talk to computers. On one end of the circuit, an operator orders his computer to send data addressed to someone else by radio up to *PACSAT*. The satellite receives the data and holds it in memory. When an operator at a distant location wants to read his electronic mail sometime later, he

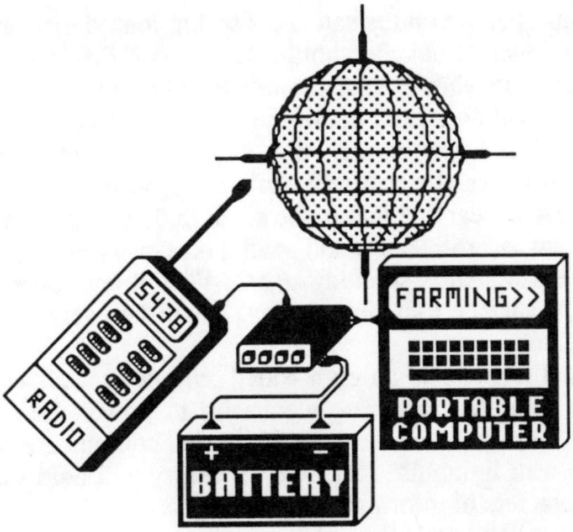

PACSAT Earth station...

tells his computer to send up a go-ahead order. The satellite, determining that the intended recipient is on hand on the ground, takes the information from its memory bank and sends it down to Earth.

The operator on the far end, after reading his mail, can compose a reply if he wishes and send it up to the satellite. The satellite will hold the mail electronically in its memory until the addressee sends up a radio signal. Then the satellite transmits the reply to the ground.

The radio signals are called packets because information being transmitted is separated into groups of letters, numbers and spaces. A packet usually contains 128 characters of information. Each set of 128 characters is sent to or from the satellite by separate radio transmission. A complete message may contain many packets totaling thousands of characters. A computer in the satellite and computers at the ground stations sort out the packets so they read in proper order.

Mailbox in a gas can

A ground station, including computer, software, radio transmitter, antenna and a special packet controller, can be purchased over the counter today for about $1,000. All will accept battery power so they will work where no power lines exist. All will fit in a suitcase.

PACSAT will fit in a NASA space shuttle Get Away Special or *GAS can.* Shaped like a garbage can, the container is 28 inches high and 22 inches in diameter. For $10,000, NASA rents the can for a ride to Earth orbit. For another $10,000 the *GAS can* user gets an optional flip-open lid on the can and a spring-loaded ejector to push the satellite away from the shuttle. Of course, *PACSAT's* trip to space depends on when NASA's shuttle fleet flies again.

PACSAT will be kicked out of the shuttle at an altitude of about 500 miles. That's low enough that simple whip antennas, made from coat hangers if necessary, will be good enough for Earth station radios to hear and talk to the satellite. It will spin around the globe, passing over the north and south poles every hour and a half. Since the Earth also is spinning, *PACSAT* will pass over any one point on the surface four times a day. It will be overhead about a dozen minutes.

The satellite will be an eight-sided cylinder coated with solar cells. The cells will generate electricity to power the satellite's computer, transmitter and receiver. The power consumed would light a weak 30-watt lightbulb. *PACSAT's* memory will hold some four million characters of information at a time.

Ten magazine articles

When the satellite is overhead, a ground station can send up messages at the rate of nearly 500 characters each second. During the entire time overhead, a ground station might be able to send up a total of 200,000 characters of information—equivalent to ten magazine articles. Mail, if picked up immediately by the recipient on his next available overhead pass, will be in the satellite from a few seconds to more than ten hours. Or it can stay in the satellite for days, waiting for radio commands from the addressee to show up.

A busy operator at a remote location on the ground will be able to program his computer to determine when the satellite will be overhead, turn on the radio gear and send up a signal asking if any messages are on hand. If messages for that operator are stored in the satellite, the computer can order them sent down. The on-the-ground computer then can store them for future reading. The ground computer can turn off the radio gear as the satellite passes out of sight over the horizon.

VITA is a non-profit group interested in sending useful non-military information to developing countries. VITA has a large library of data and pays for research into farming systems, windmills and efficient stoves and ovens. Information from VITA's library goes to its volunteer field workers who usually are in remote isolated locales.

AMSAT and VITA started the *PACSAT* low-orbit computer mailbox project in 1982. Amateur radio operators, who are building the satellite with volunteer time and donated money, also will use it, once it is carried to orbit by a shuttle, for casual contacts with hams around the world. A similar packet-radio mailbox is already in space in the orbiting *Fuji OSCAR 12* spacecraft built and operated by Japanese amateur radio operators.

Rockets

The U.S. Rocket Inventory

Shuttle grounded. Titan grounded. Atlas grounded. Delta grounded. Ariane grounded.

The year 1986 saw U.S. and European capacity for carrying men and satellites to orbit mostly grounded. With man's insatiable drive to the stars requiring space exploration, with man's insatiable drive for profits fueling commercialization of space, ways to get there were—and are—needed.

China talked to customers about future launches from Asia. Japan sent up small satellites, quietly developing a lift capability for the '90s. Even Russia lurched sideways into capitalism, saying it would take anybody's satellite to space for $43 million.

As NASA and the European Space Agency rush to get their ferry services running again, the question is: what rockets are available to carry U.S. machines and men to orbit?

There are seven readily-available ways for the U.S. to get to space: the reusable shuttle and six throwaway rockets. Here's how they stack up:

Shuttle. Designed to be used a hundred times each, three remain on hand but are grounded until safety improvements can be made. NASA may receive money to build a fourth. The 184-ft.

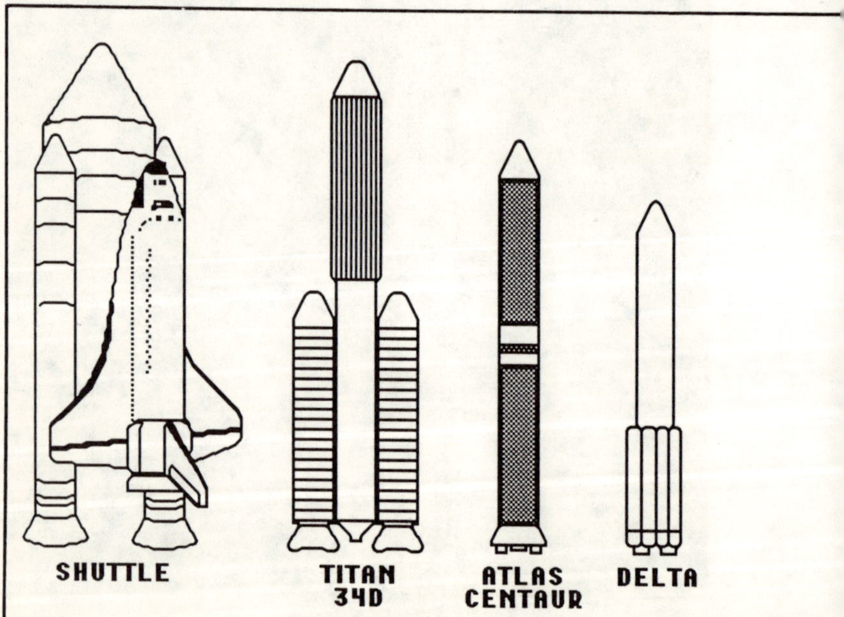

shuttle can carry 58,100 lbs. to low-earth orbit but safety modifications may reduce that slightly.

Titan 34D. Six are on the shelf but they, too, are grounded after mishaps. The 161-ft. rocket can ferry 32,900 lbs. to low earth orbit.

Atlas Centaur. This 137-ft. rocket is okay to fly but the U.S. only has two in inventory. It can lift 5,100 lbs. to a high orbit which places the satellite 19,000 miles out. From there the satellite can be transferred to stationary orbit at 22,000 miles.

Delta. Eight of these 116-ft. boosters are on hand. Each can ferry 6,350 lbs. to low-earth orbit or 2,850 lbs. to a high orbit for transfer to a stationary orbit. Delta was grounded last year but is okay to fly now.

Scout. As many as 21 of these may be on hand. The 75-ft. rocket can launch only 570 lbs. to low-earth orbit.

Atlas EF. The U.S. has four of these 68-ft. boosters. Each can carry 3,600 lbs into low-earth polar orbit from the West Coast.

Atlas. There are 15 of these in inventory to carry 3,000 to 13,500 lbs. each to low-earth orbit. Atlas was grounded last year but now is okay to fly.

A low-earth orbit would be about 100 nautical miles out in space.

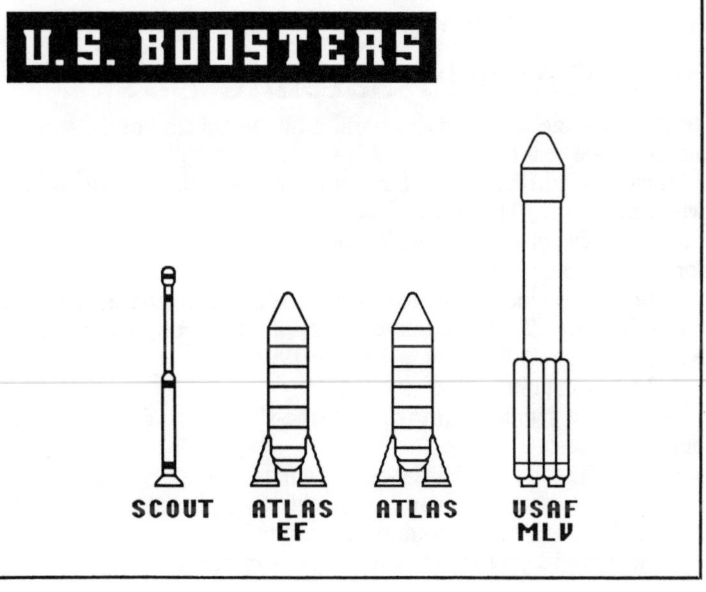

Medium Vehicle. When shuttle *Challenger* exploded, the U.S. Air Force was left with 21 navigation satellites to put in space and no way to get them there. Now they have assigned McDonnell Douglas Corp. the task of building 20 MLVs—medium launch vehicles—to carry the Navstar global positioning satellites to orbit by 1991.

Navstar satellites will be used by soldiers, tanks, aircraft and ships for precise location fixes anywhere in the world. Soldiers in the field will carry 15-lb. radios which can be tuned to the satellites. Private automobiles, trucks, boats and planes also are expected to use the navigation satellites.

The MLV will be a slightly modified Delta. Delta, manufactured by McDonnell Douglas, has been successful in 166 blastoffs in the last 26 years. It has been about 98 percent successful in the past nine years. The first model 30 years ago was called Thor-Delta and carried a 100 lb. payload.

The 20-rocket fleet will cost $734 million. McDonnell-Douglas now won't have to spend its own money to develop a rocket suitable for carrying medium-weight commercial cargo to space. Once the MLV proves itself in flight, McDonnell will compete with Martin Marietta Corp. which makes the Titan rocket and with Europe's Ariane rocket for commercial payloads.

Martin Marietta has a contract with Federal Express, the overnight package delivery firm, to launch a satellite with a Titan in 1989.

Generic Satellite Bus

NASA is getting ready to build a bus to haul robot passengers around space near Earth.

The bus will be a small generic satellite and the passengers won't be people. The bus will be a platform on which researchers can mount and change many different kinds of scientific experiments for use in orbit.

The first science package to ride as a passenger on the satellite bus will be *EUVE*, the Extreme Ultraviolet Explorer, a small space telescope to study the stars in 1990 or 1991.

The bus is built in three sections:

✔ a basic satellite built from off-the-shelf parts already used in other satellites;

✔ a data section with solar-power wings and tape recorders;

✔ and a platform to carry different payloads to be changed in orbit by shuttle or space station astronauts.

The bus will get to space via shuttle or expendable rocket.

Last Commercial *Delta* Launch

After 26 years, NASA used its last *Delta* rocket for commercial flights March 20, 1987, when it launched Indonesia's *Palapa B2-P* to orbit from Cape Canaveral.

Three remaining *Deltas* in inventory will go for military Star Wars test flights. NASA has only two more flights scheduled in 1987, both military payloads: one on *Delta* and one on *Atlas-Centaur*.

An *Atlas-Centaur* rocket failed in March 1987 when NASA launched it into a thunderstorm. Lightning struck the nosecone, wrecking its internal guidance system and causing ground controllers to destroy it.

However, the U.S. space agency had two successful *Delta* shots in less than a month. *GOES 7,* the East Coast weather satellite, was sent to orbit from Cape Canaveral February 26, 1987, aboard a *Delta.* And *Palapa B2-P* was successful March 20, 1987.

B2-P, second of three spacecraft in the *Palapa B* series, is a relay station in space capable of handling 24,000 simultaneous telephone calls to 165 million Indonesians on that nation's 3,100-mile-long archipelago of 13,677 islands, and to other southwest Pacific nations. It adds telephone, television, telegraph and data communications channels for Indonesia and the Association of South East Asian Nations (ASEAN). ASEAN includes the Philippines, Thailand, Malaysia, Singapore and Papua, New Guinea. Communications with many islands is possible only through satellites.

B2-P is in stationary orbit some 22,300 miles over the southwest Pacific. *Palapa B2-P* was to have been ferried to space aboard an American shuttle in 1986 but that plan was dropped after the *Challenger* disaster. It became the first payload to be switched from a shuttle to a non-reusable rocket after *Challenger*.

The change was made after Indonesia's foreign minister made a personal request during President Reagan's visit to Indonesia in May 1986. The foreign minister told President Reagan Indonesia wanted to replace an aging satellite already in orbit before elections. The U.S. state department persuaded the Pentagon to delay a Star Wars test flight so the Indonesians could use the *Delta* rocket. The satellite was launched March 20 and on April 23, 1987, Indonesians elected members of their national House of Representatives. The U.S. Stars Wars launch was rescheduled for November 1987.

Indonesia paid $50 million for the rocket flight in addition to $43 million for the satellite and $19.3 million for flight insurance.

Palapa B1 was ferried to orbit by a shuttle in June 1983. *B2* was carried to orbit, along with Western Union's *Westar VI* satellite, by shuttle *Challenger* in February 1984. They were dropped off in orbits lower than planned so both satellites failed. They were retrieved by shuttle in November 1984 and brought down to Earth.

More Delta rockets could be manufactured in the U.S. in the future.

U.S. Titan 3B Works

A U.S. *Titan 3B* rocket boosted a U.S. Air Force radio-ear-in-the-sky spy satellite to orbit 100 miles above Earth on February 11, 1987.

The launch from Vandenberg Air Force Base in California left the payload in a polar orbit which takes a satellite over the USSR. It was reported an *Agena* rocket is attached to the satellite. The *Agena* will push the communications reconnaissance satellite out to an orbit ranging from 1,000 miles to 22,000 miles above Earth.

Titan 3B, smaller version of the powerful *Titan 34D*, still is one of the largest rockets available to the U.S. military. It can carry 7,000 lbs. to a 100 mile high orbit. The secret satellite ferried to orbit February 11 may be able to pick up radio signals from foreign countries.

A *Titan 34D* exploded in an August 1985 launch when fuel leaked and a fuel pump didn't work. In April 1986, another *Titan 34D* blew up at the launch pad. Flame burned through insulation on a booster rocket strapped to the Titan and damaged the pad. All *Titans* then were grounded until the pad damage and rocket problems could be fixed.

Lightning Struck Atlas-Centaur

A four-bolt lightning blast punched a hole in an Atlas-Centaur nosecone as the rocket rose through a cloud bank above Cape Canaveral March 26, 1987, driving the huge vehicle into a sharp right turn, forcing NASA ground controllers to explode it.

Four lightning strikes were recorded near the rocket launch pad less than a minute after the rocket lifted off and seconds before it broke apart. A 2-ft. charred segment of fiberglass and cork nosecone with a pinhole burned in it was recovered. The hole in the fiberglass was as large as a quarter.

NASA, criticized for blasting off in severe weather, said the launch team didn't violate rules, including one that liftoff not occur if lightning is within five miles of the pad. A lightning hit was recorded 16 minutes before liftoff about 16 miles from the pad. About 48 seconds after launch, four lightning strikes were recorded within 1.9 miles of the launch area. More than 2,000 lightning flashes were striking Florida when the rocket was exploded.

The composite material in the 30-ft.-long, 10-ft.-diameter nosecone is known to build up static electricity and produce sparks when flying at high speed through rain.

The $78 million Atlas-Centaur rocket was carrying an $83 million *Fltsatcom F-6* military communications satellite. The nosecone was in place to protect the communications satellite and would have been opened like a clamshell at high altitude. NASA recovered the unfired explosive bolts which would have blown away the nosecone in space.

The rocket went out of control 51 seconds into the flight at an altitude of 14,250 ft., only one-half mile downrange from launch complex 36B, as it passed through the rain clouds. NASA destroyed the 137-ft. vehicle. Debris fell on parts of Cape Canaveral Air Force Station as well as in the ocean. More than 100 pieces have been recovered.

A NASA videotape showed a bolt of lightning darting from where the rocket was hidden in clouds and striking near the launch pad. Critics recalled that bad weather at launch time may have contributed to the shuttle *Challenger* disaster in 1986. NASA apparently knew machines flying through storms could create lightning. An F-106 jet used by NASA for tests was hit 700 times with lightning between 1980 and 1986. A researcher said that lightning actually was triggered by the presence of the plane.

However, in 2,000 launches from the Cape over 30 years, only one other rocket is known to have been hit by lightning. A mammoth *Saturn V* bound for the Moon in 1969 with three men in the *Apollo 12* capsule took a bolt which disrupted power. The astronauts fixed the problem and went on their way.

The U.S. space agency had two successful *Delta* rocket launches in less than a month. *GOES 7*, the East Coast weather satellite, was sent to orbit from Cape Canaveral February 26. And the Indonesian satellite *Palapa B2-P* was successful March 20.

NASA has only three more flights scheduled in 1987, all military payloads: one on the medium-size *Delta* rocket, one on another large *Atlas-Centaur* rocket and one on the smaller *Scout* rocket.

Fltsatcom F-6 was second in a series of three second-generation communications satellites to orbit for the Navy. The first was launched in December 1986. The satellites are shared with the Air Force and other Defense Department users. The Navy's *Fltsatcom F-8* communications satellite is scheduled for a ride to space May 21, 1987, on another *Atlas-Centaur* from Cape Canaveral. *F-8* was to have completed the series of three for the Navy.

Sometime in September, 1987, a *Scout* rocket fired from Vandenberg Air Force Base in California will carry a pair of *Oscar* navigation satellites to orbit for the Navy. The package to be launched is called *SOOS-2*, for Stacked Oscar On Scout. *SOOS-1* went to space in August 1985. These *Oscar* satellites are not the same as the better-known *Orbital Satellites Carrying Amateur Radio (OSCAR)* communications relays in space.

Sometime between October, 1987, and December, 1987, NASA will use a *Delta* rocket from Cape Canaveral to launch a Strategic Defense Initiative Star Wars payload. That will be the second of four *Delta* launches of SDI space weapons.

Rocket Explodes Over Alaska

A *Black Brandt 10* sounding rocket failed seconds after launch from NASA's Poker Flats, Alaska, launch site in March 1987.

The lost payload had been a plasma physics experiment to be carried out in the upper reaches of Earth's atmosphere on the fringe of space.

Black Brandt is a solid-fuel rocket. Its second stage failed about ten seconds after ignition and the rocket exploded at 20 seconds into the flight.

Heavy-Lift Rocket Planned

NASA wants to help the Air Force develop a new heavy-lift launch vehicle (HLLV) to carry space station parts to orbit in 1994.

The Air Force needs a powerful rocket to carry heavy Stars Wars military weapons to space in the 1990s. NASA needs a heavy lifter to haul construction materials for the space station to orbit.

An HLLV could carry more weight than the space shuttle, reducing the number of flights required per year for the shuttle fleet in the 1990s.

An HLLV might carry in one flight a complete station module which would require two or three shuttle flights.

Once the station is built and permanently manned, NASA plans to rotate crews eight times a year so it must fly up to the station every 45 days. However, used equipment and trash to be carried back down to Earth from the station may require more than eight shuttle flights a year.

NASA is designing an all-new solid-fuel booster rocket for use on shuttles after 1992. It will replace the booster style which caused the 1986 *Challenger* disaster.

Manufacturers will compete for a construction contract to be let in 1989. No shuttles have flown since January 1986.

No Centaurs In Shuttles

A powerful rocket fueled with liquid hydrogen and liquid oxygen has been ruled off the U.S. space shuttle.

NASA decided the *Centaur* rocket is too dangerous to carry into space aboard the shuttle. *Centaur's* fuel is the same combination which destroyed *Challenger* on January 28, 1986.

Centaur was to have blasted off from an orbiting shuttle in May 1986, sending a probe to orbit Jupiter and another past Jupiter to the Sun.

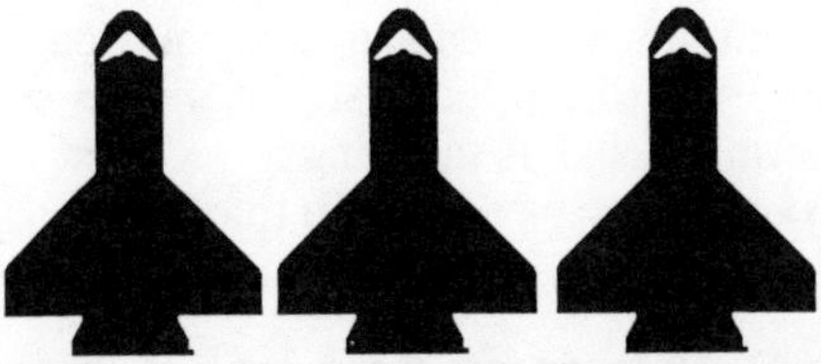

The shuttle version of *Centaur* is an adaptation of a rocket which has been used for ten years as a final stage atop unmanned rockets. It sent *Viking* to Mars. A future shuttle *Centaur* would have sent a probe to radar-map Venus. Others would have carried secret military heavyweight satellites.

The various missions will go forward but not until *Centaur* has been rebuilt for use atop unmanned rockets. The U.S. Air Force has ordered unmanned *Titan* rockets which could carry *Centaurs*. NASA may design a powerful shuttle-carried rocket using solid fuel.

Local Train To Space

Like a local freight train dropping off boxcars as it makes it rounds, a refurbished *Saturn* moon rocket may be able to launch six satellites at one time to different orbits.

The U.S. Air Force asked two major rocket builders to dust off plans for the old launcher which sent men to the Moon between 1969 and 1972. A 200-foot-tall rocket with a cargo area bigger than a 747 jumbo jet passenger compartment, the huge new *Jarvis* medium launch vehicle (MLV) would ferry six satellites to various orbits ranging from low-Earth about 500 miles altitude to geostationary orbit 22,000 miles out in space.

The proposed MLV is named after engineer Gregory B. Jarvis who died in the *Challenger* shuttle disaster in January 1986.

The Air Force wants the *Jarvis* MLV for use in the 1990s. Its nosecone will be powered so it can fly to different altitudes. Sidewalls of the nosecone will open like flower petals. The six satellites will be pushed off into space as the nosecone passes through different orbits. The MLV will not be reusable.

Small Van To Cart Stuff Down

NASA is looking for designs for a small reusable free-flying space hauler to be ferried to space with its own scientific payload to fly alone back down to Earth later.

A 600 to 800 lb. package would be loaded into the *free flier* and it would be sent to orbit in the space shuttle or atop a non-reusable rocket such as the Delta.

When its payload was no longer needed in space, the flier would fire a braking rocket and drop down into the lower atmosphere for a soft landing on Earth with no damage to the payload.

NASA may want to use the first free flier in space as early as 1988. West Germany also is planning a free flier. Animals used in experiments could be one payload for a free flier.

First Commercial Launch

The Martin Marietta Corp. may be the first private U.S. corporation to make a commercial launch when it sends up a Federal Express Corp. satellite in 1989.

The U.S. government is encouraging private firms to fill the launch void created when NASA space shuttle flights were restricted to military and scientific satellites.

Who Needs A Small Satellite?

All of a sudden private rocket launch companies are ready to haul stuff to space. But who needs their services?

For $15 million, Space Services Inc. (SSI) of Houston will bolt your 3,000-lb. satellite to its romantically-named *Conestoga* rocket and cart it out to low-Earth orbit for you to use.

Or, for a mere $8 million, American Rocket Company (Amroc) will do the same, using its not-so-cleverly-named *Industrial Launch Vehicle (ILV)*.

One problem may be, nobody has a small satellite needing a lift to orbit right now. Where's the market? The government could be one customer. And news organizations who would like to have their own spy satellites up there. And oil companies along with other users of Earth's natural resources. Possibly the U.S. Defense Dept. or the CIA. Maybe even NASA itself.

Meanwhile, Amroc hasn't even flown its rocket. The *ILV* has been tested on the ground at the U.S. Air Force Rocket Propulsion Laboratory. Amroc would like to blast one downrange on a suborbital trip in 1987 to see if it works. If it does, they might try to go for orbit in 1988. The Defense Dept, is likely to put a secret payload atop the *ILV* if it goes to space in 1988.

Houston Industries Inc., a local power company, recently sunk millions into SSI in the hope that private launches of commercial satellites will catch on. A Canadian investment bank is raising money for Amroc.

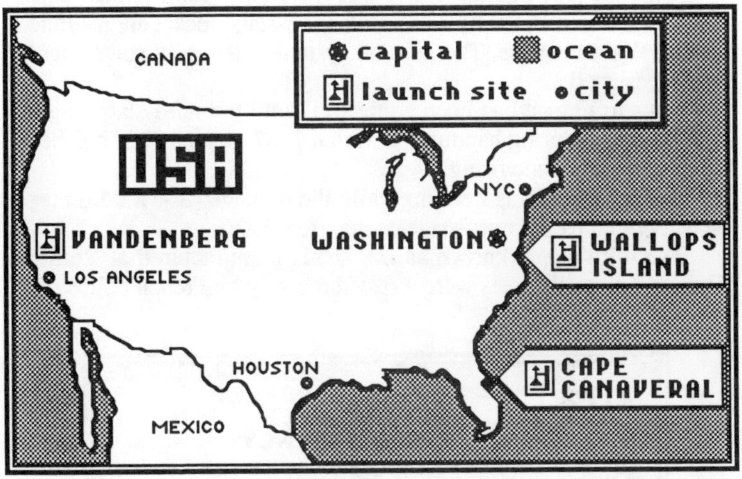

Space Rocket Names

There are many space rocket names. Here's a pretty complete list:

•The United States has Scout, Delta, Delta 2, Atlas, Atlas G, Atlas-Centaur, Titan 2, 3, 4, and 34D, and shuttles Discovery, Atlantis and Columbia.

•The USSR has SL-13, SL-4, Vertikal, and Proton.

•European Space Agency uses Ariane 2, 3 and 4.

•Japan has H-1A, H-2, N-2 and Mu-3S-2.

•India uses versions of the SLV-3.

•People's Republic of China has FB-1, CSL-2, CZ-1, CZ-2 and CZ-3.

Additional rockets are being tested and their names will be heard in the future.

Satellite Insurance

The satellite insurance industry paid more than $700 million in loss claims from 1984 to 1986 while taking in $325 million in premiums. Rates charged satellite owners were doubled after a string of space mishaps. Some owners paid $10 million to insure a $40 million satellite.

European Launch Site

European Space Agency fires its *Ariane* rockets to space from the Kourou launch site in French Guiana, northeastern South America.

Ariane V-19, in June 1987, was the first *Ariane* launch since May 1986 when *V-18* exploded as its third-stage rocket engine failed during flight to space. The ignition system in the third-stage engine was redesigned.

Tests in France had been going well until the third-stage engine was damaged during handling in March 1987. *Ariane V-19* had been set for an early-June firing.

The schedule was held up while the damaged *V-19* third stage was replaced by the third-stage engine from *V-20*.

The third stage, known as *HM7B*, is manufactured at Vernon, France, where six spies were arrested recently by French police. The

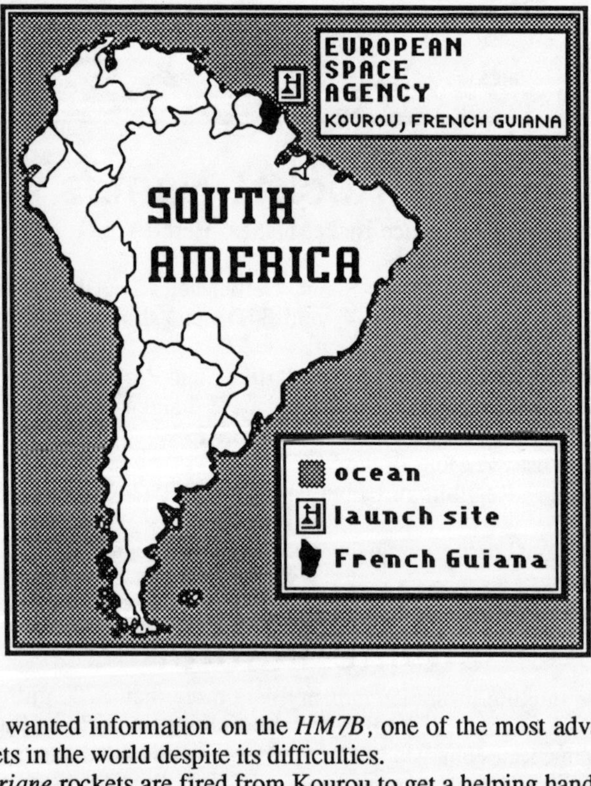

spies wanted information on the *HM7B*, one of the most advanced rockets in the world despite its difficulties.

Ariane rockets are fired from Kourou to get a helping hand from Mother Nature. Since the site is almost on the equator, where Earth rotates fastest, less booster power is needed to push a satellite to Earth orbit. Satellites are launched north and east from Kourou. The tropical site has not had a launch delayed by bad weather.

Ariane Grounded

The European Space Agency grounded its *Ariane* payload launcher fleet in 1986 and 1987 for investigation of a blast-off failure.

Three rocket models, *Ariane 2, 3* and *4*, were affected. An *Ariane 2* failed in May 1986 triggering the investigation. An *Ariane 3* failed in September 1986. The new *Ariane 4* is untried.

The U.S. shuttle disaster and failure of two other major U.S. launchers in 1986 had persuaded some satellite owners to use *Ariane* to get to orbit. When *Ariane* also was grounded, some asked China and other spacefaring nations for reliable launching rockets.

The shuttle disaster was January 28, 1986. An Air Force Titan rocket exploded April 18, 1986. NASA's Delta rocket failed May 3, 1986. The Ariane failed May 30, 1986.

Russia Tests Huge Rocket

The USSR test fired, on the ground at the Baikonur Cosmodrome at Tyuratam, a huge new heavy-lifter rocket capable of ferrying a 220,0000-lb. payload to orbit.

The booster, tied to its pad for the test, delivers 6.6 million-lbs. of thrust, similar to *Saturn V* which lifted Americans to the Moon in 1969-72. An unmanned test flight is expected in the summer of 1987. A manned flight with the new Russian shuttle attached could come in 1987 or early in 1988.

The U.S. abandoned *Saturn V* a decade ago with the decision to go for a reusable space shuttle. A new U.S. heavy-lifter rocket on the drawing board for mid to late 1990s will lift only 100,000 to 150,000 lbs. to orbit.

The USSR shuttle has no built-in rocket motors. Rather, the main engines are in the bottom of the large rocket core. Four liquid-fuel booster rockets are strapped around the core. A shuttle or an unmanned cargo container can be strapped to the core.

Cheap Lifts

The USSR announced early in 1987 it would make inexpensive space flights available to other countries needing satellites ferried to orbit.

To make it easy for other nations to use their new profit-making service, the Russians will exempt payloads from customs inspection when they enter the USSR. Payloads will be permitted to be shipped in sealed trucks to the Baikonur Cosmodrome near Tyuratam in the Central Asian desert.

A worldwide shortage of launch vehicles developed in 1986 when U.S. and European Space Agency boosters were put on hold after failures. The Chinese immediately started offering commercial rides on their *Long March* series of rockets. The Russians have followed suit.

Three USSR boosters will be used for commercial launches—the heavy-lifting *Proton* which has flown successfully on 90 out of 97 tries since 1970; the *SL-4* which the USSR uses to launch its own *Soyuz* manned spacecraft and *Progress* cargo ships; and a smaller rocket known as *Vertical* which has been used for atmospheric soundings with recovery capsules. *Vertical* can provide commercial space research service short of carrying large cargos to be dropped off in orbit.

The USSR will provide insurance coverage from its own government insurance company. Flights on behalf of developing countries in Asia, Africa and Latin America will be carried at price discounts.

To profit from their past space efforts, the Russians will start renting out channels in their *Gorizont* communications satellites. And they plan to sell data about resources on Earth collected by various satellites they have in orbit. They will develop custom experiments and manufacturing systems for use aboard USSR spacecraft on behalf of commercial firms around the world.

Proton rocket. The superstar of their commercial launch business currently is the *Proton* rocket. It can carry 20 tons of goods to orbit 125 miles above Earth. Or it can ferry two tons of payload to stationary orbit some 22,000 miles out in space.

Main competitors for *Proton* are the European Space Agency's *Ariane* rockets and the Chinese *Long March* launchers. The U.S. space shuttle travels only to low Earth orbit and then only with medium-weight payloads. U.S. expendable rockets have been in very short supply. NASA planned only six launches of expendable rockets in 1987.

Europe's satellite telecommunications organization is considering switching to *Proton* to launch its 3,680 lb. *Eutelsat 2*. Under consideration as backups are two U.S. rockets, *Titan* and *Atlas/Centaur*. The European organization had planned to ferry *Eutelsat 2* to low Earth orbit aboard a European Space Agency *Ariane*. The U.S. space shuttle was to have been the backup carrier. But, after NASA's 1986 *Challenger* disaster, *Eutelsat* was among several commercial satellites bumped from future shuttle rides. The Russians are said to have quoted a price to carry *Eutelsat 2* to orbit about half what the *Ariane* ride would cost.

The USSR has said it will send 20 tons of payload, aboard *Proton*, to stationary orbit for $43 million payable in Swiss francs.

Three of the *Eutelsat 2* satellites are under construction and eventually will need rides to space. The first had been planned for launch in the summer of 1989 on *Ariane*. The European organization previously had committed themselves firmly to three *Ariane* launches and one U.S. shuttle launch, plus two optional rides on *Ariane* and one on the shuttle. Now *Ariane* and the shuttles are grounded and NASA bumped *Eutelsat* in favor of military and scientific satellites.

The Russians told the European organization they could launch the first *Eutelsat 2* for $26 million with only two years lead time. In the event of *Proton* failure, another *Proton* rocket could be ready to go in four weeks if another satellite were ready for it.

Proton has three rocket stages when flying to low orbit. At the bottom of *Proton*, the first stage is a central fuel tank with six booster rockets strapped to its sides. A second stage is bolted above the first stage and a third stage is mounted atop the second stage. A fourth stage is added when a payload must be carried 22,000 miles out to stationary orbit, or on interplanetary flight.

Unlike U.S. technique, USSR rockets are assembled while resting horizontally on their sides. They are transported by railway in that position from assembly building to launch pad where they are erected to a vertical position for blast off.

The USSR probably will send nearly 100 satellites to orbit again in 1987, most to be named *Cosmos*. The number compares with six 1987 launches planned by NASA.

Numbers designating different Cosmos satellites are approaching 2000. Names of other Russian satellites likely to be heard include *Raduga, Molniya, Meteor, Gorizont* and *Radiosputnik.* Two amateur radio satellites in the *Radiosputnik (RS)* series are expected to be carried to orbit.

Shuttle. The USSR may be able to carry satellites and other cargo in a new space shuttle expected to be test-fired by early 1988.

And a huge new heavy-lifter rocket capable of carrying more than 110 tons to orbit was test fired successfully in May 1987 from Baikonur Cosmodrome. The booster delivers 6.6 million-lbs. of thrust, similar to *Saturn V* which lifted Americans to the Moon in 1969-72. A manned flight with the new Russian shuttle attached could come late in 1987 or early in 1988.

Superficially, the Russian space shuttle looks like the U.S. shuttle. But, unlike the U.S. shuttle, the USSR shuttle has no built-in rocket motors. Rather, the main engines are in the bottom of the huge rocket core. Four liquid-fuel booster rockets are strapped around the core. A shuttle or an unmanned cargo container can be strapped to the core for flight to orbit.

The powerful new 100-ton lifter also could be used to carry giant freight loads to low-Earth orbits 500 miles high, heavy-weight satellites to stationary orbits at 22,000 miles out, or men and materials to the Moon or Mars.

The USSR's space shuttle already has been mounted for ground testing on the side of the huge rocket on its new launch pad beside a 3.5-mile shuttle landing runway, roads, buildings and railways at the Baikonur Cosmodrome in the eastern part of the the USSR state of Kazakhstan.

The USSR also may be preparing a smaller space plane to fly only people to orbit and back, as well as a larger space truck like the U.S. shuttle to carry both men and cargo.

H-2 Rocket Due In 1991

Japan is expected to start using a powerful new launch vehicle, the *H-2* rocket, in 1991 to ferry satellites to Earth orbit.

Technology in the current *H-1* rocket was borrowed from the United States. Agreements between the two countries prevent Japan from seeking commercial launch contracts for *H-1* flights. The *H-2* will not be so encumbered and Japan is expected to be competitive in the space-launch business when the *H-2* is ready.

Also, Japan probably will build a small reusable space shuttle to be launched atop the *H-2* rocket. A larger shuttle could be built later and launched on an *H-2* with extra booster rockets strapped to its sides.

China Improving Launch Site

China is expecting to do a lot of satellite launch business in future years. They are building a second launch pad and a building where satellites can be checked and repaired at their Xichang launch complex. They're also building a new hotel for visiting businessmen.

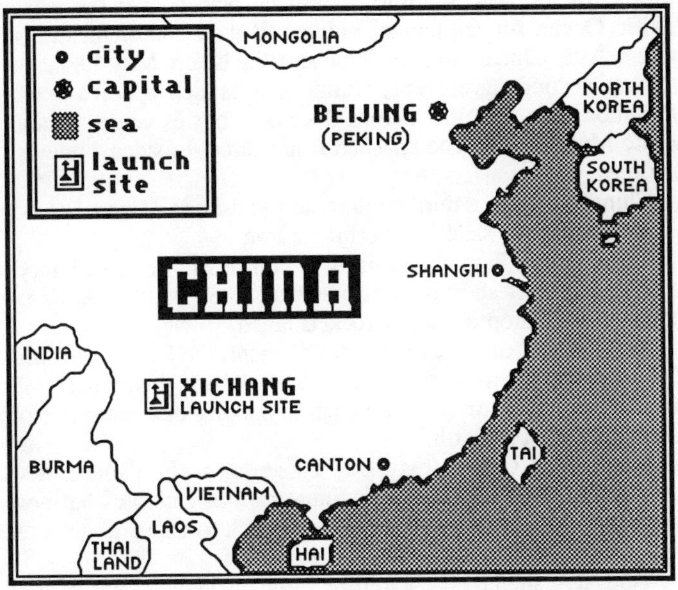

Long March Of Men Into Space

Chinese astronauts will be launched to orbit soon, according to *People's Daily,* the official newspaper.

The first flight is "not far off," the paper said in 1986. Life-support systems and a crew cabin reportedly are ready to fly atop a modified version of China's reliable *Long March* workhorse rocket.

China has been working on bringing spacecraft back to Earth safely for ten years. Spacesuits and food have been in the works for seven years. Astronauts training in a mock spaceship first showed up in *Science Life,* a Chinese magazine, in 1980. A first-flight crew is being selected now, according to the paper. A Chinese shuttle may be on drawing boards.

Four satellites are booked to fly on the People's Republic of China *Long March* rocket in 1987 and 1988.

The Great Wall Industry Corp., a government body which schedules Chinese space flights, has signed contracts to launch two television broadcast satellites and two communications satellites.

One satellite, formerly known as *Westar 6*, didn't work after being carried to orbit in a shuttle and had to brought back down on a later flight. It was fixed and renamed *Pacific Star 1*.

The new owner of *Pacific Star 1* paid insurance companies $20 million for the used satellite. It will be placed over the western Pacific Ocean for telephone, video and data communications in eastern Asia. China will launch the satellite before May 1988.

In a second agreement, China will launch a private, U.S., Dominion Video satellite before December of this year and another before March 1988. The spacecraft are direct-broadcast television satellites.

China agreed in a third contract to use a *Long March 3* to boost a communications satellite to orbit for Iran.

The People's Republic is using its various *Long March* models to fill a launcher shortage left when the U.S. shuttle *Challenger* exploded and European *Ariane* rockets failed.

Another part of the Chinese government, the People's Insurance Co. of China, will underwrite a $60 million insurance policy covering *Pacific Star 1* from launch to parking in a stationary orbit 22,000 miles from Earth.

Meanwhile, Canada may take a *Long March* to space. NASA launched Canada's first nine communications satellites but now it looks like Canada will make the long march to China to get a tenth bird in orbit about 1990.

Telesat Canada is a satellite builder owned jointly by the Canadian government and private industry. Rocketry in China is run by a government agency, the Ministry of Astronautics.

With the U.S. shuttle fleet grounded temporarily and a worldwide shortage of boost-to-orbit flights, Telesat has been talking with Great Wall about using China's *Long March* rocket in 1990.

And there may be a *Long March* for Sweden. A Swedish spaceflight group has contracted with China to use the *Long March* rocket to boost a commercial satellite called *Mail Star* to orbit.

Mail Star will permit Earth stations on one side of the globe to send typewritten messages up for storage aboard the satellite. Stations on the other side of the globe can command *Mail Star* to send down the messages when the satellite is overhead.

Other known customers using *Long March* rockets include Western Union and Terasat.

There had been controversy over the possibility China could steal space technology when satellites were shipped there for launch. But the Chinese government customs department has agreed not to inspect the satellites. Satellite owners will be permitted to carry their own hardware to the launch site in western China. A permit from the U.S. State Dept.'s Office of Munitions Control is required before a satellite can be shipped to China.

Indian Satellite Launch Failure

India's satellite launch attempt failed March 24, 1987, when its *Augmented Satellite Launch Vehicle* fell into the Bay of Bengal less than one minute after liftoff from the Sriharikota Space Center in the southeastern state of Tamil Nadu.

The five-stage 40-ton 77-ft. rocket was carrying a 300-lb. weather and communications satellite.

Smaller versions of the rocket which failed have placed 80-lb. satellites in orbit. India, which had seven payloads in orbit at the end of 1986, builds many of its own. India uses liquid fuels while most satellites today are launched with solid-fuel rockets.

The rocket type which failed will be part of a larger launcher to

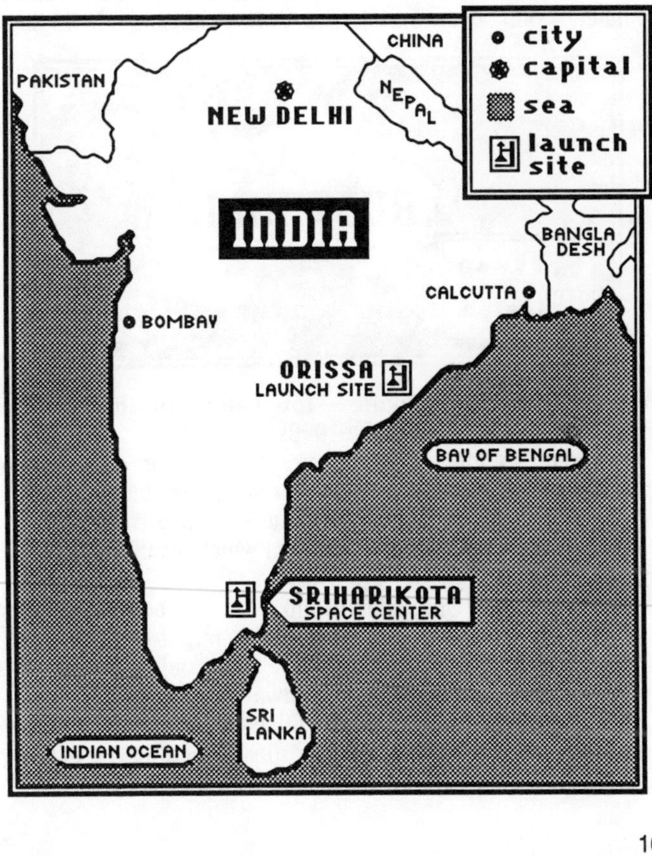

carry 2,000-lb. satellites to orbit in the 1990s. The failure was the third among five attempts since 1979.

A giant *Insat* communications satellite, designed in India and launched earlier by the U.S., provides a national tv link, long-distance telephones, and meteorological reports for the country which is two-thirds agricultural and dependent on weather patterns.

Another *Insat* was to have been launched on a U.S. shuttle but now will be carried aloft in 1989 by a European Space Agency Ariane rocket.

India also is building a military rocket range in the eastern state of Orissa.

Spaceport For Rent

If you're going to be needing a spaceport in the next decade, you might want to consider something near Weipa, Australia.

The Queensland state government there has been putting a lot of thought into trading the excellent climate in the remote, sparsely-populated 4,000-square-mile Cape York Peninsula—including its

overpopulation of crocodiles—for a slice of the worldwide spaceflight launching pie in the 1990s.

It's certainly a low-overhead area at the moment. And it probably can guarantee longer life in space for satellites since it's about 15 degrees below the equator. A satellite launched from Cape York Peninsula would need 120 lbs. less fuel which means a couple years more life in space.

The premier of Queensland thinks the weather at Weipa is as near perfect as one could want. That might be attractive to governments and private companies who could replace Mother Nature with steel towers, concrete runways, and hotels.

A Queensland University scientists says spaceport operators would earn millions of dollars in profits, even after the expense of moving the crocodiles elsewhere.

Solar
System

Exciting '90s: On To Planets, Asteroids, Comets, And Sun

The United States has sent more than 50 automated spacecraft over 30 years into the vast emptiness between planets of our Solar System. These sophisticated robots were the first to explore the Moon, the planets, the Sun. The 1990's are going to be even more exciting for space watchers with exploring machines scheduled to leave for Mars, Jupiter, Venus, the Sun, the Asteroid Belt and comets. Here's where we're going:

Jupiter

Galileo is a billion-dollar mothership which will fly five years to Jupiter, orbit the planet, and drop a science probe which will parachute into the colossal planet's stormy atmosphere.

The *Galileo* mission to Jupiter is scheduled tentatively to be launched from a U.S. shuttle in 1989. It was to have been launched from the shuttle *Atlantis* payload bay in May 1986 using a hydrogen-fueled *Centaur* rocket. The January 28, 1986, *Challenger* explosion led to concerns of danger and cancellation of NASA plans to fire *Centaur* from the shuttle. The 1989 launch will use a low-power inertial upper stage (IUS) solid-fuel booster rocket from the shuttle cargo bay.

Galileo is heavy and the IUS booster is not as strong as *Centaur*. So, NASA is planning a slingshot trip for the spacecraft. *Galileo* will be taken into near-Earth orbit aboard a shuttle. The IUS will be fired toward Venus. The craft will whip around Venus and back toward Earth. The gravity pull encountered in the Earth flyby at only 800 miles altitude will fling *Galileo* at high speed toward Jupiter. The five-year trip would have required only two years with a direct flight on a *Centaur* rocket. NASA expects the added time for flight controllers and data analysis to cost an extra $50 million a year.

The Sun

The *Ulysses* spacecraft, designed to study the Sun at close range, also was to have been launched in May 1986 with a *Centaur* fired from the shuttle *Challenger* cargo hold. Now it will be sent out in 1990 from a shuttle using the 17-foot-long IUS booster rocket.

Ulysses, constructed by the European Space Agency, will orbit the Sun's poles. To get there it will first travel to Jupiter. The giant planet's gravity will whip *Ulysses* above the plane of the Solar System from where it will dive back toward the Sun. There are no rockets with sufficient power to send Ulysses on such a route directly from Earth.

The IUS rocket has a good track record. It was used three times in earlier shuttle flights. Once it launched from the shuttle a NASA

Tracking and Data Relay Satellite (TDRS) and twice secret military satellites. An IUS booster and a *TDRS* were in *Challenger* when the shuttle exploded.

A powerful new *Titan 4* non-reusable rocket may be available in 1989. NASA has the option of switching one of its interplanetary craft to *Titan* for launch.

Venus

The *Magellan* spacecraft will orbit Venus, using radar to create maps of the planet surface. It may be aboard the third shuttle flight in 1989. The IUS would boost it along an interplanetary trajectory.

Asteroids

NASA is planning an interplanetary spacecraft to test the Solar System's Asteroid Belt on its way for a close-up look at Comet Tempel-2.

The *Mariner Mark II* will have a camera and a dozen scientific instruments when it leaves Earth in 1992 in search of clues to the origins of our Universe.

Mariner Mark II will go beyond the orbit of Mars to that dangerous area of big rock chunks known as the Asteroid Belt. It will be controlled from Earth to fly close to the asteroids Hestig and Malautra.

Asteroids are rocks, boulders and tiny planets travelling together in orbit around the Sun, filling a ring between the orbits of Mars and Jupiter. The Belt includes about 1500 rocky objects ranging in size from boulders up to minor planets of about 300 miles diameter.

Comets

Passing through the zone of asteroids, *Mariner Mark II* will travel on out to rendezvous with the comet Tempel-2 in 1996. The comet, floating along its orbital path, takes between five and six years to circle the Sun. Mariner Mark II will be ordered from Earth to slow down and fly alongside Tempel-2 for three years.

Monitors in the spacecraft will analyze dust, gas and plasma around the comet as it nears the Sun and its tail grows. Information will be sent by radio to Earth.

About midway through their tandem travel, in 1997, *Mariner Mark II* will be commanded to fire a probe filled with scientific instruments three feet deep into the nucleus of Tempel-2.

Scientists say they hope to uncover clues about the beginning of the Solar System and the start of life on Earth.

Mars

The *Mars Observer* spacecraft is the U.S. entry in the race Mars. It will be launched in 1992 and will reach the red planet four years after a planned Soviet orbiter, leaving in 1988, gets there. However, *Observer* will fly closer to the planet's surface.

The USSR craft will stand off some 4,000 miles from the

surface while *Observer* will be stationed about 250 miles above Mars. The US last studied Mars with the *Viking* orbiters and landers back in 1976.

Observer will have a moving-mirror infrared eye to see what's in the dust in Mars' atmosphere as well as on its surface. A sharp camera will shoot the best pictures ever taken of Mars from orbit. The camera includes a compact telescope. Pictures will show two-mile squares of the Martian landscape. *Observer* will have a radar altimeter to map hills and valleys on the surface and electronics to read Mars' magnetic field. A gamma-ray scanner will look for uranium, iron, potassium, calcium, magnesium and other key elements.

Observer will be in orbit around Mars and will not land on the surface. It will catalog the planet's atmosphere, especially its famous dust. And the satellite will spot different kinds of rocks on the surface.

Neptune

Continuing its magnificent tour of the Solar System, *Voyager 2* will sail within 3,000 miles of the planet Neptune on August 24, 1989.

NASA scientists had wanted to send the spacecraft within 800 miles of the planet's surface but decided against it when they realized Neptune's orbiting rings and atmosphere might risk damage to the probe.

Voyager will take man's first close-up photos of Neptune. In 1986 the spacecraft provided exciting scientific data when it sailed past Uranus.

Neptune is second most distant of the Sun's nine major planets. Uranus is third most distant. Billions of years ago, our star—the Sun—exploded into life at the core of a rotating disk of hot dust and gas. The nine planets and other pieces of our Solar System formed out of the swirling materials in the outlying regions of the disk. The planet Mercury is closest to the Sun. Venus is second and Earth is the third planet from the Sun. Mars is fourth followed by Jupiter, Saturn, Uranus, Neptune and Pluto.

Some astronomers suspect there may be other planets, beyond Pluto, yet to be discovered by man. The planets are so distant it will take man a year or more just to visit our closest neighbors, Mars and Venus, and return. It is thought men will go to Mars within the next 20 years.

Voyager 2 is controlled by NASA's Jet Propulsion Laboratory at Pasadena, CA.

Space shuttle

The next flight of the U.S. space shuttle is scheduled for sometime in 1988. *Discovery* is scheduled for the first flight of the shuttles after the 1986 *Challenger* disaster. At least three other shuttle systems are under consideration elsewhere in the world.

The USSR is completing a shuttlecraft. The Europeans are drawing up plans for a *Hermes* shuttle with final decision to go forward in 1987. The Japanese are studying whether to build their own shuttle for use in the mid-1990s.

Space station

The U.S. and other nations in the 1990s will join the USSR as countries with space stations in orbit around Earth. The USSR has had space stations in orbit since the early 1970s. Its second-generation *Mir* station is now in use. The U.S. had a station for a short time in the mid-1970s. The United States second-generation station is expected in 1995 or later. The U.S. has no station in orbit now.

The Europeans are planning *Columbus,* a manned module which can be attached to the U.S. space station in the 1990s or could be expanded to become a separate European station. The Japanese also are building a module for the U.S. station.

Deep space telescope

The fifth launch of the U.S. shuttle in 1988 may be of great importance to mankind. It is expected to carry to Earth orbit the 2.4-meter *Hubble Space Telescope.* That instrument will see light no man has seen before, coming from stars so far away they sent out their light during the very earliest times of the Universe.

The telescope, which was to have gone to orbit aboard a shuttle in 1986, will be used for at least 20 years. Astronauts in shuttles will make service calls from time to time to keep the telescope in good working order. NASA has called the *Hubble Space Telescope* "the greatest scientific instrument that man has ever built."

More Sunspots On The Way

Our star, that flaming gasbag we affectionately call Ol' Sol, will be showing a lot more internal turmoil through the 1990's.

Solar activity varies on a regular schedule covering eleven years on man's calendar. This so-called *Sunspot Cycle* regularly affects radio, television and even household electricity on Earth.

Our Sun is a large, hot, spherical star nearly a million miles across. That's 109 times bigger than planet Earth making the Sun our prime source of light, heat and energy. Earth and other planets in our Solar System revolve around the Sun.

Astronomers looking at the Sun discovered a long time ago that sometimes the face is pockmarked. The blemishes are *sunspots.*

A sunspot is a cooler spot on the Sun's hot surface. It looks like a patch on the face of the Sun, a small light ring with a dark center.

As the Sun rotates, a sunspot will seem to move across the face and disappear behind the Sun. If it has not abated in the meantime, a sunspot will reappear 27 days later when the Sun completes one rotation.

Sunspots appear and disappear. Their number reaches a maximum every eleven years. Magnetic storms and disruptions of radio communications on Earth are associated with large numbers of sunspots.

More sunspots mean more solar activity going on. More activity in the Sun means improved worldwide radio transmission on Earth. Here's why:

The Sun. Our star is 864,948.7 miles in diameter. Compare that to Earth's diameter of 7926.7 miles. The surface of the Sun is about 6,000° Celsius. Inside it increases to 13 million degrees. That internal temperature is hot enough for nuclear fusion to take place. Hydrogen is converted to helium and tremendous amounts of energy are given off. The energy is released as electromagnetic radiation of all wavelengths, including radio and television frequencies.

Brief energy outbursts are seen on the Sun's surface as bright *solar flares*. Solar flares give off extra strong radio waves as well as protons and electrons.

The Sun radiates in all directions and much energy arrives at Earth, only 93 million miles distant. Most of the energy here is absorbed by our atmosphere.

The Ionosphere. In the highest reaches of Earth's atmosphere, near space, beyond our surface weather, is a layer filled with clouds of electrically-charged particles called ions. These floating areas of ions form the ionosphere. Radiated energy from the Sun charges the particles in the ionosphere. More sunlight means more ions and a more thoroughly reflecting mirror for radio signals.

A radio signal moves around the world bouncing between the ionosphere and Earth's surface. Originating at a transmitter, the radio signal travels up to the ionosphere. The ionosphere acts like a mirror to the radio wave, reflecting it back down to Earth. The surface bounces it up again and the ionosphere reflects it back down. The radio signal seems to skip around the globe. Listeners with receivers along the path can detect the signal even if they are on the far side of the Earth.

Yesterday, today, tomorrow. Casual radio and television users may have noticed sunspot peaks in the past. Distant television stations could have interrupted local programs. CB radios may have heard stations 1,000 miles away better than transmitters across town. Shortwave listeners would have found it easier to tune in the BBC or Radio Moscow. Police dispatchers sometimes had local calls interrupted by signals from cities hundreds of miles away. Ham radio operators found it easier to talk with amateurs in more than

200 countries around the world. Brief electrical outages on Earth have been noted at times of worst solar disturbances.

We were at the bottom or trough of the sunspot cycle in 1986. In a few years we will be at the peak of the sunspot cycle and the enhancement of radio and television signals probably will occur again. As at past peaks, astronomers will renew efforts to study the Sun. The most active year probably will be 1991.

It's not known why more sunspots appear during some cycles and fewer at other times. One of the most active peaks in recent memory was during the 1957-58 high. Flat cycles with little enhancement to radio transmissions have been recorded. Scientists can't predict for sure which type we will experience through the 1990's.

SOLAR-A To Study Flares

To study solar flares from a better vantage point, Japanese and U.S. scientists will place an eye in the sky called *SOLAR-A*.

The craft will be launched into a low orbit, near Earth, in 1991.

SOLAR-A will be one of a series of satellites to be launched by Japan from their Kagoshima Space Center aboard *M-3S-II* rockets between August and September 1991. That period will be known for highest solar activity during the current Sunspot Cycle.

SOLAR-A, flying at an altitude of about 344 miles, will stay in orbit for three years. During that time, it will stare straight at the Sun, looking for and examining solar flares on the face of our star.

U.S. NASA scientists will supply the optics and detector for a soft x-ray telescope aboard *SOLAR-A*. The Japanese will build the electronics, computer and power system for the satellite.

The satellite will see x-rays and gamma rays coming from the Sun. Soft x-rays are weaker x-rays containing less energy.

Old Sunspots

The famous 11-year Sunspot Cycle has been going on for millions of years.

Huge chunks of ancient rock, imprinted with 100-foot-long dark lines every half inch, have been dug from the bottom of an Ice Age lake in Australia.

Over millions of years, ice in glaciers melted and the water running off carried silt to the lake bed. The marks were written on the rocks when layers of silt were laid down each spring at the bottom of the lake bed. The warmer a spring, the greater the runoff and the more silt deposited on the lake bed.

In prehistoric times, as now, the changing pattern of spots on the face of the Sun led to varied weather on Earth, changing the amount of spring runoff from land with glaciers.

Geologists have correlated the lines on the Australian rocks with modern sunspot cycle records back to 1800.

The Sun Is Less Bright

Our star has lost about one-tenth of one percent of its brightness in recent years but scientists say there's nothing to worry about.

Although the dwindling brightness could mean a drop of temperature by one-fifth to two-fifths of a degree on Earth, that apparently has been offset by the greenhouse effect around this planet.

The *Solar Maximum Satellite*—affectionately known to scientists as *Solar Max*—has been reporting on the Sun's brightness since 1980. Solar Max found about one-thousandth of the light radiated from the sun was missing by 1985.

If the solar output continued to dwindle, life on Earth would be threatened. Scientists at NASA's Jet Propulsion Laboratory at Pasadena, CA, think there is no chance of the phenomenon creating a new Ice Age on Earth because they haven't noted any further decrease since mid-1985.

Scientists don't agree on whether or not the dip in Sun brightness relates to the eleven-year sunspot cycle. The last sunspot cycle peak was in 1980 but a Nimbus 7 satellite first reported the diminishing brightness in 1978.

Sunspots are dark, cool spots on the Sun's surface where magnetism inside prevents heat from bubbling out to the surface. Solar flares, on the other hand, are extra-hot spots erupting outward on the Sun's surface. Some solar physicists think the current loss of brightness may be related more to increased solar flare activity than to sunspots.

Most say the amount of light radiated from the Sun will start to increase in 1987 as the sunspot cycle starts to build again toward a 1991 peak.

There was a period of very low sunspot activity between the years 1650 and 1725. For about 300 years, centered on the late 1600s, Earth shivered through what has come to be known as the Little Ice Age. Farming was hurt and glaciers grew southward when temperatures dropped two to three degrees Fahrenheit around the world.

If the 1980s dimming continues for another 50 years, another Little Ice Age could result.

However, the current loss appears to have been offset by the greenhouse effect. Environmental pollution gases from our industrial society warm the Earth by trapping heat like glass panes in a greenhouse.

The news from *Solar Max* has disturbed scientists who had been laboring under the belief that the Sun would become more dim as more cool sunspots appeared on its face. Now astrophysicists say the Sun may be brighter during sunspot peaks because the spots are accompanied by bright hot spots called faculae or "little torches."

Downturns and upturns in solar output, as measured by satellites like *Solar Max* in orbit around Earth, seem tied to decrease and increase in hot spot activity on the Sun.

Sailing To The Sun

The European Space Agency's *Ulysses* spacecraft will be launched toward the Sun from an Earth-orbiting U.S. shuttle in October 1990.

NASA worked out a schedule with ESA after *Ulysses* and the American *Galileo* interplanetary probes were delayed by the 1986 *Challenger* disaster.

Galileo will be launched from a space shuttle toward the planet Jupiter in October or November 1989. The launch dates were selected to bring maximum return of data from the two important scientific flights.

Although launched one year later than *Galileo*, the 121-lb. *Ulysses* will begin to send back prime data in 1994. That's one year earlier than *Galileo* which has a longer trip time.

Delaying *Galileo* to 1991, the next available launch window, would have brought a two-year delay in studying the space environment around Jupiter. NASA decided that trying to launch both *Ulysses* and *Galileo* within a three-month period in 1989 would be a risky rush.

The 11-ft.-long *Ulysses* originally was known as *International Solar Polar Mission (ISPM).* It will be first to explore what is called the "third dimension" of the solar system by flying high over the Sun's poles.

It carries nine instruments to measure magnetic field, plasma, electrons, protons, ions, interstellar neutral gas, gravity waves, solar particles, cosmic rays, plasma waves, radio signals, x-rays, gamma rays and cosmic dust.

Some 120 scientists from 44 labs in 12 countries are involved in the experiments.

69mph Balloon Studies Sun

An unmanned balloon on a leisurely float from Alice Springs, outback in Australia, caught some rays for a couple of weeks in February 1987 as it floated around the globe at 69 miles per hour.

NASA's balloon, cruising above 100,000 miles, carried instruments to see small hard x-ray flares and super-hot flare plasma on the Sun. When it slowed to 10 mph over Brazil and started down, NASA ended the flight February 21.

An earlier balloon was blown down by a thunderstorm over Paraguay February 6. When the balloon wandered into cool air, it dropped below 60,000 feet. A pressure sensor pushed the payload overboard for a parachute ride to the ground.

The second balloon was launched February 9, 1987, to circumnavigate the globe. Neither balloon made the complete trip.

NASA hasn't scheduled another balloon but the sighting of a nearby supernova probably will lead to another flight.

Dust Rings Around The Sun

The Sun is circled by huge rings of interplanetary dust.

One ring of dust, between Jupiter and Mars, has been known for a long time. It causes what astronomers call Zodiacal Light.

Another ring was discovered by the Infrared Astronomy Satellite (IRAS) in 1983 among the asteroids between Mars and Jupiter.

Japanese and Indonesian astronomers found a third ring in 1983 near the Sun. The dust in that ring looks like it has been spiralling slowly inward from the outer solar system. The Sun vaporizes it and the gas is blown back outward from the Sun by the pressure of solar radiation.

Sun WIMPs

A Swiss scientist is blaming WIMPs for a shortage of neutrinos coming from the Sun.

Neutrinos, tiny parts of atoms, overflow from the Sun and stream outward. Some of these particles, landing on Earth, are found by physicists who think there should be three times as many arriving here as are counted.

To explain the mysterious shortage of neutrinos from the Sun, a California astrophysicist designed the WIMP theory.

A WIMP is a Weakly Interacting Massive Particle found, in his theory, at the center of our star. It increases the density of matter in the Sun and increases the gas pressure by finding extra hydrogen in the Sun's heart.

Without substituting big particles like WIMPs for the missing small neutrino particles, scientists would have to lower their estimate of the temperature in the middle of the Sun. A lower temperature, however, would mean lower gas pressure inside the Sun and it wouldn't be able to hold up its outer layers.

Now a Swiss scientist at the World Radiation Center has found what may be proof that there really are WIMPs in the Sun.

Without WIMPs the density of stuff in the Sun's core is about 166 grams of matter per cubic centimeter. If there are WIMPs, density could be as high as 196 grams per cc. That's about 196 times as dense as water on Earth.

Pioneer 9 Declared Dead

Pioneer 9 spent 20 years checking on solar wind and cosmic rays in a gigantic orbit around the Sun. But it was 1983 when NASA last heard from the spacecraft so they've given up trying.

The 148-lb. spacecraft was launched November 8, 1968, with the expectation its data would beep back for only six months or so. Surprising controllers on Earth, it kept sending back information until May 18, 1983. NASA engineers at Ames Research Center, Mountain View, CA, have been trying since then to talk *Pioneer 9* into answering calls from home—with no luck.

On March 3, 1987, they sent out one last set of 270 radio instructions and stood by their most-sensitive receivers. One receiver used is so sensitive NASA usually dedicates it to the search for extraterrestrial intelligence. Nothing was heard. *Pioneer 9* was given up for dead.

Seven still work. Meanwhile, seven other *Pioneers* still are on the job. *Pioneers 6, 7* and *8*—among the earliest interplanetary probes—are doing jobs similar to *Pioneer 9's* assignment. *Pioneers 10* and *11* are speeding out of the Solar System after making important calls in the vicinity of Jupiter and Saturn. *Pioneer 12* is in orbit around Venus.

An interplanetary probe is a spacecraft which flies away from Earth, beyond the Moon, to or near planets and other bodies in our Solar System. The *Pioneers* orbiting the Sun were the first to take a close look at the solar wind, solar magnetic fields and cosmic rays. Scientists think the Sun is typical of many stars in the Universe so *Pioneer* data has been useful in understanding how stars work.

Until 1972, *Pioneers* supplied data on storms on the Sun which can affect power and radio signals on Earth. Unfortunately, those four *Pioneers* haven't been tracked much in recent years since newer spacecraft have stolen operator time on NASA's limited number of Deep Space Network antennas.

Pioneer 9 is like a tiny man-made planet. It circles the Sun inside Earth's orbit, having gone around the Sun 22 times to date, travelling a total of 11 billion miles. It takes 297 days to go around the Sun, passing from within 70 million miles to 90 million miles of the Sun. The small satellite sent back 4.25 billion bits of data while it still could be heard.

What happened? Engineers guessed a part wore out inside *Pioneer 9* leading to an electrical short-circuit. Or the craft may have been hit by a meteorite.

Pioneers are powered by solar cells. As they have aged, all four *Pioneers* orbiting the Sun have turned themselves off to save power when farthest from the Sun. Usually, mission controllers have been able to turn them back on by radio command. Not so with *Pioneer 9* since 1983. *Pioneer 9* will continue to loop around the Sun indefinitely.

Before *Pioneer 9* took a closer look, solar wind was thought to be a steady gentle flow of ionized gases spiraling out from the Sun in a stream rushing along at a million miles per hour. Instead, the spacecraft encountered a region of great turbulence with fast-moving streams ploughing through slower ones, creating shock waves. Even though the matter distributed through space is in a vacuum far better than any created on Earth, scientists still found interplanetary space to be an exciting place.

Pioneer 9 measured the twisted magnetic fields threading the solar wind and the streams of high-energy particles which follow the course of the magnetic field out from the Sun.

Boiling wind. It studied cosmic rays to see how many there were and what type they were, uncovering how they vary with the cycles of the Sun. Scientists now understand more about the Sun's corona from which the gases of solar wind boil off into interplanetary space.

When it was behind the Sun from our vantage point here on Earth, *Pioneer* helped predict storms on the Sun. It saw and reported events on the backside of the Sun two weeks before they could be seen on Earth. Huge bursts of solar wind buffet and distort Earth's magnetic field, throwing circuit breakers in power systems and causing power black-outs. Storms on the Sun even may cause some of Earth's long-term weather. In the United States, the Federal Aviation Administration (FAA), the military, power companies and radio and tv stations benefitted from *Pioneer's* predictions of solar storms.

Dangerous protons. When astronauts took their *Apollo* spacecraft to the Moon, the *Pioneer* spacecraft sent mission control hourly reports of activity on the surface of the Sun. The information helped guard against an unexpected arrival at the Moon of intense showers of dangerous protons from the Sun.

In August 1972 during the most intense solar storm ever recorded, *Pioneer 9* luckily was aligned with *Pioneer 10,* 132 million miles distant on its way to Jupiter. The two craft measured the gas from the Sun upstream and downstream. Scientists found the gases flowing away from the Sun slowed to half speed but got a lot hotter. The solar wind was converting much of its motion energy to thermal energy. *Pioneer 9* measured the highest solar wind speeds ever known.

Pioneer designs were simple yet rugged. *Pioneers 10* and *11* were the first spacecraft to pass through the Asteroid Belt and reach Jupiter. Now they are heading out to interstellar space from opposite sides of the Solar System.

Mercury Atmosphere

There appears to be far more sodium than anything else in the thin atmosphere of the planet Mercury.

Pioneer Switches To Wilson

NASA's shopworn but highly serviceable *Pioneer 12* spacecraft, which has been peering down at Venus from orbit around that planet since 1978, temporarily turned its attention to Comet Wilson, discovered in 1986 by a Cal Tech graduate student.

The reliable old satellite, which already has studied three other comets while on duty near Venus, spent April 1987 measuring the rate of water evaporation from Comet Wilson. The body is named after Christine Wilson, the California Institute of Technology PhD candidate who discovered it in August 1986.

Pioneer also will see how much carbon and oxygen the comet may be giving off. It will do that by looking at the comet in the ultraviolet portion of the electromagnetic energy spectrum.

Wilson is streaking past Venus in an arc around the Sun. Its closest approach, called perihelion, will be April 20. *Pioneer 12* will spend 20 hours a day studying Wilson while it is nearby.

Fresh comet. Wilson is unusual for observers—a fresh comet embarking on its first Solar System visit. Fresh comets are of special interest because they have not been close enough to the Sun before to change their characteristics. Such virgin comets may provide a better record of the primitive conditions of the early Solar System millions of years ago.

Comets probably are remnants of the dust cloud which became our Solar System. Planets and other objects condensed. Dust and gas which became comets may have been driven to the outer reaches of the Solar System by pressure of sunlight shortly after the Sun and planets formed.

Comets are mostly water ice, dust and rocks, sort of celestial fossils which may reveal unique information from the past. In fact, comets also may carry a record of activity in the outer areas of the Solar System. For instance, cosmic rays from our Milky Way galaxy may have worked on the outer surfaces of comets. Conditions in the galaxy in times past may be read from comet outer layers.

A large cloud of comets—the Oort Cloud— may be lurking just beyond our sight, circling the Sun, billions of miles away at the outer edges of our Solar System.

Shock wave. Maybe, several million years ago, some passing star came close enough to our own Solar System to perturb this vast array of comets. This shock wave threw some comets out of their conventional orbits into highly elliptical paths which periodically bring them near the Sun.

Pioneer 12 studied Comet Encke in 1984, Comet Giacobini-Zinner in 1985 and Halley's Comet in 1986. Encke is thought to be near the end of its life so *Pioneer 12* will study it in June and July 1987 when Encke comes close again.

Meanwhile, *Pioneer* will try to learn something about Wilson's skin of carbon-bearing molecules. Fresh comets have such a skin which melts away under close exposure to the Sun. Comparing carbon evaporation with water evaporation, scientists will see if Wilson has such a skin of carbon molecules.

Water bursts. If there are outbursts of extra carbon or extra water, scientists will find the comet's nucleus is patchy, not uniform. If the nucleus is not smooth, they will be able to deduce the rate of rotation of the nucleus by timing the outbursts of carbon and water.

Pioneer 12 is situated well to study Comet Wilson. Ultraviolet measurements cannot be made on Earth since the ozone layer in our atmosphere blocks most ultraviolet light. Along with *Pioneer*, spacecraft in orbit around Earth will help in the study of Wilson. The *International Ultraviolet Explorer IUE* satellite in Earth orbit will scan the comet, looking for other chemicals. Since *IUE's* prime mission is looking at more distant stars and galaxies, not much of its time will be spent on Wilson.

Pioneer 12 has been ordered to turn 10 degrees so it can see the comet.

U.S. and USSR To Cooperate

If U.S. President Reagan and USSR General Secretary Gorbachev can get together in 1987 they probably will sign a space-cooperation agreement.

The pact will bring cooperation in unmanned exploration of Mars, other planets and deep space.

Reportedly the agreement includes coordinated efforts between the U.S. and USSR in 16 space programs. Data would be exchanged before and after those experiments. The two countries would not build joint hardware.

The USSR wants to cooperate in the late 1990's on an unmanned mission to bring soil samples from Mars to Earth for analysis.

In 1988 or soon after, the USSR will send a craft to study Phobos, the larger and innermost of two natural satellites of the planet Mars. The U.S. *Mars Observer* spacecraft is planned for 1992. The two countries would share the scientific results of those flights under the proposed pact.

Similarly, the U.S. is planning to send a *Magellan* spacecraft to study Venus. Data from that and USSR probes to Venus would be shared.

The countries also would share information about Earth-bound studies in astronomy, astrophysics, space sciences, earth sciences and life sciences as they apply to space.

A 1972 agreement between the two countries led to the joint Apollo-Soyuz flight test. Unfortunately, that agreement was allowed to expire in 1982 after the Russians used troops to enforce martial law in Poland. The new agreement will be more specific than the 1972 document.

Japan: To The Moon In 1990

Japan will send a scientific explorer around the Moon in 1990, dropping off a miniature lunar satellite on the swing-by.

Moving to display a higher profile in space, Japan plans to fire the *Muses-A* dual satellite toward the Moon to practice for future interplanetary spaceflights.

Shaped as a 430-lb. 5-ft. diameter 3-ft. tall cylinder with a 26-lb. 1-ft. miniature satellite attached to one end, *Muses-A* is expected to leave during a 20-day launch window in January-February 1990 aboard an MU-3S2 booster for one year of maneuvers in space between Earth and the Moon. As *Muses-A* swings around the Moon on the first of several fly-bys, it will drop its baby satellite into lunar orbit.

Muses-A will look at bright stars and the Moon's edge to navigate its course. If all goes well, the spacecraft will encounter the Moon, release its tiny companion satellite, and swing by within 14,100 miles of the Moon's surface on March 20, 1990.

On board will be a West German micrometeorite counter from the Munich Technical University which will record the weight, speed and direction of dust particles striking the craft.

By its fifth fly-by on August 6, 1990, *Muses-A* will be travelling out more than 600,000 miles from Earth. Japanese scientists expect to maintain a one million kilometer apogee distance for the rest of the spacecraft's life.

Muses-A, among other jobs, will be a test flight for a future mission called Geotail, part of an international Sun-Earth physics research project.

The mini orbiter released by *Muses-A* uses up 11 of its 26 lbs. weight in a retro motor it needs to move into orbit around the Moon. The polyhedral-shaped orbiter is covered with solar cells for electricity. A tiny cross-shaped antenna protruding from the top of the lunar orbiter allows direct contact with its controllers on Earth.

Growing Moon Food

Will wheat and soybeans grown in dirt on the moon?

Minnesota scientists have powdered about 200 pounds of rock quarried near Duluth. They say it's just like lunar soil. They even put in some glass.

The light porous glass in moon dirt comes from the heat of

countless meteor impacts over eons so they used a plasma-arc furnace to make some. The furnace also will boil off moisture in the Earth rock making it more like lunar soil.

The stuff has been trucked to Florida's Epcot Center where wheat and soybean seeds have been planted in it. If they grow, moon residents 20 years from now may need only a greenhouse with atmosphere to feed themselves.

Moon's Helium Dust

An odd kind of helium dust, which NASA thought was a useless find on the Moon, may make nuclear power plants on Earth safer.

Geologists think there may be one million tons of the rare Helium-3 in moon dust. That would supply the world's needs for hundreds, even thousands of years.

Electricity has been produced by nuclear fission in which atoms are split. Now some scientists want to switch to nuclear fusion where atoms are joined together.

Fusion plants supposedly would be safer. If they used Helium-3 for fuel they would be even safer since waste from the plant would be so low in radioactivity it could be buried near the Earth's surface.

To get the fuel, a mining station would have to be placed in space to mine the moon's surface, refine the Helium-3, and ship it back to Earth. Proponents say all that effort would still be ten times as efficient as present uranium energy yields.

Scientists had known about the fusion possibilities for Helium-3. They had discounted it as impractical since Helium-3 is scarce on Earth. *Apollo* moon missions uncovered the rich deposits.

First Camera On The Moon

The first snapshot camera carried by American astronauts on the Moon was European, originally designed in 1948 in Goteborg, Sweden, by Hasselblad.

The first one had an erratic focal plane shutter and some American-made Kodak Ektar lenses. It made negatives about 2-1/4 inches square, larger than the popular 35mm format.

The astronauts didn't like the shutter and lenses so leaf shutters were substituted and Zeiss lenses were added. NASA also adapted a motorized Hasselblad 500EL for lunar surface photography.

They built a frame viewfinder so astronauts could compose pictures through a spacesuit helmet. They added an extra-large latch so astronauts could change film magazines while wearing bulky spacesuit gloves.

Birth Of The Moon

Our Moon may have resulted from a mid-Space collision between our planet Earth and some other huge planetary body eons ago when our Solar System was forming.

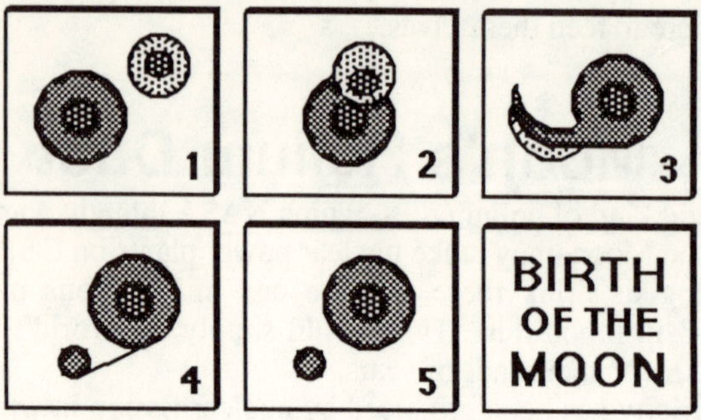

The mystery planet, about one-seventh the size of Earth, approached ❶ and they collided ❷.

Heat from the explosive impact melted the two planets' iron cores together and threw chunks of mantle from both bodies off into space ❸.

The iron core remained as Earth while the weight of the spun-off debris led its gravity to clump it into a Moon-shaped ball ❹.

The new theory from New Mexico researchers explains differences in amounts of chemicals found in the separate Moon ❺ and why the moon has only a tiny iron core.

The Race To Mars

The U.S. entry in the race to expand knowledge of Mars will be the *Observer*, a probe NASA hopes to launch in 1992.

Mars Observer will reach the red planet two years after a planned Soviet orbiter gets there in 1988. However, *Observer* may fly closer to the planet's surface.

In 1988, the USSR will send a craft to orbit Mars and a second to the Martian moon Phobos. The Mars craft will stand off some 4,000 miles from the planet surface while the later U.S. *Observer* will be stationed about 250 miles above Mars.

NASA plans for *Observer* to depart for Mars from an Earth-orbiting shuttle. With shuttle mission plans up in the air, critics have asked NASA to bolt *Observer* atop an unmanned expendable rocket for a 1990 launch.

The U.S. last studied Mars with the *Viking* orbiters and landers back in 1976. Key questions about Mars remain. Federal budgets for space research have been tight. To save money, *Observer* will be a redesigned communications satellite with some weather satellite parts added on. It will be small, lightweight with limited data handling capability. Yet its few sophisticated instruments will be expected to add to our knowledge of Mars.

Included will be a moving-mirror infrared eye to see what's in the dust in Mars' atmosphere as well as on its surface and a sharp camera for the best pictures ever taken of Mars from orbit. The camera includes a compact telescope. Pictures will show two-mile squares of the Martian landscape.

Observer will have a radar altimeter to map hills and valleys on the surface and electronics to read Mars' magnetic field. A gamma-ray scanner will look for uranium, iron, potassium, calcium, magnesium and other key elements.

In orbit around Mars, *Observer* will catalog the planet's atmosphere, especially its famous dust. And the satellite will spot different kinds of rocks on the surface.

USSR Plans Mars Explorers

The Russians have an energetic plan to send high-tech robot probes every two to four years to explore Mars in the 1990's:

•1988, the USSR will send a spacecraft to orbit Mars with cameras and other instruments. And they will send a second spacecraft to orbit Mars' moon Phobos with cameras and other instruments.

•1992, the USSR will send a more intricate Mars spacecraft including a small Mars-rover vehicle, camera-carrying French atmospheric balloons, ground penetrators, cameras, a soil analysis lab and a map maker. The 500-lb. package of instruments will include tv, radar and analysis equipment plus two robot moles to dig into the Martian soil. One balloon will be released to bounce and drift over the surface of Mars at an altitude of three to four miles during the day as the sun heats the balloon. It will take high-resolution pictures of what it sees as it ranges across 3,000 miles of

Martian landscape. When the balloon sinks to ground at night, its penetrator will analyze the chemical composition of surface soil and matter deep in the ground. It also will look for evidence of life in the form of organic material in the soil.

•1994, the USSR will send a Mars craft with a larger ground rover vehicle and instruments. Two robot moles will be able penetrate the Martian soil to a depth of 100 feet. The larger rover will have a greater range than the 1992 vehicle.

•1996, the USSR will send a spacecraft with a large rover, instruments, and an ability to pick up rock and soil samples and return them to Earth.

•1998, the USSR may send a craft beyond Mars to the Asteroid Belt to land on an asteroid and return samples of an asteroid to Earth.

It's not known when the USSR plans a manned mission to Mars. A trip would take two to three years while cosmonauts so far have spent only 237 days in space.

U.S. May Try For Mars In 1998

NASA's Johnson Space Center in Houston, Texas, and Jet Propulsion Laboratory in Pasadena, California, are working out a scenario for U.S. robot spacecraft to leave Earth for Mars in 1998 and return here in 2001 with rock samples.

The spacecraft would land on Mars, rove around for sightseeing, and bring back 11 lbs. of soil, rock and atmosphere samples to Earth.

Such an unmanned flight would be a forerunner to manned visits to Mars later, NASA said.

Jet Propulsion Lab is looking into a Mars-orbiting craft and a surface rover equipped with scoops and tongs. Both would rely on their own computer programs for most work since remote-control radio signals would take an hour to get from Earth to Mars.

JPL managed the Viking lander flights to Mars in 1975 and 1976 as well as most other American interplanetary spaceflights.

Johnson Space Center is considering a spacecraft capable of flying down into the Martian atmosphere and slowing down for a gentle landing. The Martian atmosphere is only one one-hundredth as dense as Earth's atmosphere.

Johnson also is looking into the lift-off equipment to blast the craft back to Mars orbit for rendezvous with the spacecraft before return to Earth.

Returned Martian samples would be dropped off at the U.S. space station for lab analysis, similar to the chemical tests applied to Moon rocks nearly 20 years ago.

Windmills On Mars

Mars colonists in 30 years may heat their houses and operate scientific equipment with electricity from windmills.

The U.S. *Viking* spacecraft, landing on Mars a decade ago, found 15 mile per hour winds on the surface. Hills and valleys, to channel the wind naturally to man-made windmills, can be seen from satellites orbiting Mars.

Windmills were used first on Earth 1,400 years ago in the 7th Century. In the 21st Century, men may hook turbines to modern windmills, called *giromills,* to generate electricity for water and oxygen generating stations.

Photos sent back to Earth in the 1970s showed sand dunes and wind erosion, suggesting winds on Mars even stronger than 15 miles per hour.

Martian Dust Devils

Scientists have been dusting off and re-examining those 10-year-old photos of the surface of Mars made by the Viking spacecraft in orbit around Mars. What do they see? Gigantic dust devils—whirlwinds of sand and dust—towering higher than three miles above Mars' surface.

Eight Rocks From Mars

How did eight rocks get here from Mars?

Geologists, picking through 10,000 meteorites found on Earth, have wondered for years how eight of those rocks came all the way here from Mars. The most recent answer: an asteroid did it.

A meteorite is a chunk of rock now on Earth which came from somewhere else in our Solar System or beyond. The eight mystery meteorites display a chemistry similar to rocks analyzed back in 1976 when two *Viking* spacecraft landed on Mars and ran chemical tests. The eight found on Earth show a chemical makeup different from other Earth rocks.

Scientists estimate a rock on Mars would have to accelerate up to at least 11,300 miles per hour to escape the planet's gravity and travel on out into space. What would cause a rock to move so fast?

Suppose an asteroid of, say, 500 to 5000-ft. diameter were floating past Mars at, say, 16,000 miles per hour. Suppose it struck the planet a glancing blow at an angle of, say, between 25 and 60 degrees. It would be like a golfer making a long drive.

The crashing asteroid would kick up dust and gas travelling at

50,000 miles per hour. That vapor would be powerful enough to lift rocks and send them flying out and away from Mars at 11,300 mph or faster.

Sailing along through interplanetary space, some would have come close enough to be attracted down onto the face of Earth by our planet's gravity, according to astrogeologists at the California Institute of Technology.

Planet Samantha Smith

A USSR astronomer has named a newly discovered minor planet—an asteroid between Mars and Jupiter—after Samantha Smith, the U.S. child who visited Russia, became an American television star and died in a 1985 plane crash in Maine.

The late Russian leader Yuri Andropov invited Samantha to visit Russia in 1983 after she wrote him a letter in a personal campaign for better U.S.-USSR relations. Later she starred in an American television program until she died in the 1985 air mishap.

Astronomer Lyudmila Chernykh discovered the space object recently and decided to name it after Samantha. The USSR already has honored her memory with a special commemorative postage stamp, has given her name to a ship, named a new species of flower after her, and named a 12,000-foot Caucasus Mountains peak for her.

Jupiter's Colorful Clouds

The *Voyager* spacecraft, probing Jupiter's night side in 1979, for the first time spotted lightning and auroras of electrically-charged particles in the giant planet's atmosphere.

The lightning may splash Jupiter's clouds with colorful layers of blue, brown, white and red.

There are three layers of clouds 19 miles apart. Jupiter's atmosphere contains hydrogen, helium, traces of water, ammonia, methane, and organic carbon compounds.

The innermost cloud layer is water ice and droplets. The middle layer is ammonia crystals and hydrogen sulfide. The outermost layer is ammonia ice.

The blue clouds are warmest and at the lowest altitude. Brown, white, and red clouds, in that order, are higher.

Powerful lightning probably disrupts the chemicals, forcing sulfur, phosphorus, and organic compounds to color the clouds. If not lightning, it could be rapid vertical motion through changing temperature levels.

SATURN PLANET SYSTEM

Major satellites circle ringed Saturn: Dione, foreground, with Saturn rising behind. Tethys, Mimas, upper right. Enceladus, Rhea, lower left off rings. Distant Titan, top left. Phoebe, Iapetus, Hyperion, so distant they are hardly larger than star points, upper right and middle left in artist's stylization. Stats: Ten times larger than Earth, ten times farther from the Sun. Its year: 30 Earth years. A day, under eleven hours. The least dense planet, even less than water. If one had a big enough ocean, Saturn would float. Core may be rock but most of the planet is liquid hydrogen. Rings are ice chunks sized from marbles to basketballs shaped by shepherd moonlets, 50 to 100 miles in diameter. Order of satellites is Mimas, Enceladus, Tethys, Dione, Rhea, Titan, Hyperion, Iapetus and Phoebe.

Lightning On Uranus

French scientists listening to radio signals recorded in 1986 at Uranus by the *Voyager 2* spacecraft say they heard lightning bolts.

The same kind of static crashes heard in an AM radio during an electrical storm on Earth were found among data transmitted back to Earth from Uranus, the third from last known planet in our Solar System.

Apparently the lightning strikes on Uranus are more powerful than those on Earth but nowhere near as strong as the magnificent electrical discharges on Saturn. The bolts are giant sparks of static electricity generated by turbulence in the Uranian atmosphere.

When electricity flows, it generates electromagnetic radiation. This energy is what we call a radio wave. The radio wave can be controlled by man using a transmitter to generate a signal. The signal can be modulated with sounds or made to carry other forms of information. On the other hand, when the electromagnetic energy is not controlled, we call it noise or static. Lightning is merely a very large, uncontrolled burst of static.

The *Voyager* spacecraft's receiver heard strong static as it sailed past Uranus. The chaotic signals received demonstrated the same characteristics as lightning noise.

Unusual Uranus

The planet Uranus, its moons, its rings, orbiting as they do at the far reaches of our Solar System, seem even stranger as scientists analyze data received by radio from *Voyager 2* in January 1986.

The length of a Uranus day is 17.24 hours. The planet does have a magnetic field but it's tilted 60 degrees away from the axis upon which Uranus rotates making it unreadable from Earth. That tilt may mean the magnetic field is undergoing a planet-wide polarity shift.

Uranus' 15 moons are eye-openers. Statuesque mountains towering more than ten miles high. Incredibly deep valleys. Vast plains, some with a mysterious dark surface. Moon diameters from slightly fatter than 25 miles, up to about 1100 miles, circling Uranus from every eight hours out to every 14 days.

While looking at Uranus' rings, *Voyager* uncovered partial incomplete-circle formations called ring arcs. The many rings and arcs varied in depth and density.

Uranus is the third most distant planet from the Sun. Continuing its exploration of the Solar System, *Voyager 2* will sail farther out to within 3,000 miles of the planet Neptune in 1989.

Voyager 2 At Neptune In 1989

NASA wanted to hear radio signals better so it touched up the flight path of *Voyager 2* to Neptune, ordering its hydrazine thrusters to fire for about a minute and ten seconds on March 13, 1987.

The midcourse trajectory correction means the spacecraft, on the last leg of its magnificent decade-long tour of the Solar System, should fly about 3,000 miles above the planet's north pole on August 24, 1989. And NASA's biggest receiving antenna, in Australia, will be looking directly out to Neptune, listening for the flow of radio data from *Voyager 2*.

The big antenna is created electronically on Earth when NASA hooks its Deep Space Network antenna to Australia's Parkes Radio Astronomy Observatory 200 miles away. The electronic array works as an extraordinarily large and sensitive antenna.

Firing the rocket engine to speed up *Voyager* means the spacecraft will arrive over Neptune 12 hours earlier.

NASA scientists had wanted to send the spacecraft within 800 miles of the planet's surface but decided against it when they realized Neptune's orbiting rings and atmosphere might damage the probe. Neptune is thought to have unusual partial rings known as ring arcs.

Voyager 2 will make man's first close-up photos of Neptune. In 1986 the spacecraft provided exciting scientific data when it sailed past Uranus.

Neptune is second most distant of the Sun's nine major planets. Uranus is third most distant. Billions of years ago, our star—the Sun—exploded into life at the core of a rotating disk of hot dust and gas. The nine major planets and other pieces of our Solar System formed out of the swirling dust cloud in the outlying regions of the disk.

The planet Mercury is closest to the Sun. Venus is second and Earth is third from the Sun. Mars is fourth followed by Jupiter, Saturn, Uranus, Neptune and Pluto.

Some astronomers think Pluto may not be a major planet after all but they still suspect there may be other planets, beyond Pluto, yet to be discovered by man. The planets are so distant it will be a two to three year roundtrip when man finally sails off for a visit to Mars or Venus, our closest neighbors. It is thought men will go to Mars within the next 25 years.

Voyager 2 is controlled by NASA's Jet Propulsion Laboratory at Pasadena, CA. The spacecraft will be 16 times closer to Neptune that it was to Uranus, 20 times closer than it was to Saturn, and 55 times closer than *Voyager 1* was to Jupiter. While visiting in the area, *Voyager 2* will fly as close as 25,000 miles to Triton, Neptune's large moon.

Could Pluto Not Be A Planet?

Astronomers are fickle, just like everybody else.

Pluto, the ninth planet in the Solar System and farthest from the Sun, has been a planet for 57 years. But that celebrity status could change very soon now that we see the known planet farthest from Earth as the smallest in our Solar System.

Pluto has been seen only as a tiny speck moving against a background star field. We used to imagine it was big, even bigger than Jupiter which is 88,000 miles in diameter.

Unfortunately, Pluto's image has been shrinking. In fact, now it's down under one-third the size of Earth.

Pluto is a rocky ball with a metal core. It now appears to be only 1,375 miles in diameter. That's small enough to drop into the space between Dallas and San Francisco. In fact, it's almost exactly the same as the distance between New York City and Dallas.

Scientists once thought Pluto was a comet which wandered into our Solar System from deep space and was captured in an unending orbit of the Sun. But new data from German scientists studying Pluto from an observatory in Chile suggest it's more likely Pluto is a moon of Neptune escaped from the gravity clutch of that big planet.

If Pluto were a comet, it probably would be a dirty snowball splattered with chunks of rock. But astronomers now say it seems too dense for that. It probably is more like Earth's moon which is mostly rock with a small metal heart. Our moon's metal core is smaller than the metal core of Earth.

For a time, after their calculations saw Pluto shrink from 88,000 miles diameter, astronomers believed Pluto was between 1,800 and 2,200 miles in diameter. That's about the size of Earth's moon.

But as Pluto eclipsed its moon, Charon, recently, astronomers timed how long it took Charon to reappear on the other side of Pluto. From the time, they computed the distance—Pluto's diameter. It turned out to be only about 1,375 miles in diameter, much smaller than Earth's moon.

They also say Pluto has several characteristics unlike other planets in our Solar System. Thus, Pluto—named for the Greek god of the underworld—may be something less than a planet and only 1,375 miles in diameter.

Of course, Pluto is 3.6 billion miles from the Sun. At that distance, it's hard to measure. But the latest instruments picture it much smaller than we thought 57 years ago.

As things stand now, Pluto is smaller than Earth's moon and smaller than some of the moons of Jupiter and Saturn. The director of the Smithsonian Astrophysical Observatory wants Pluto officially downsized to...asteroid!

Halley's Comet Explorations

In March 1986 six spacecraft from Earth—*Giotto, Vega 1, Vega 2, Suisei, Sakigake* and *ICE*—encountered Halley's Comet at various flyby distances on different days.

Giotto, sent by the European Space Agency (ESA), was named after Italian painter Giotto Di Bondone who depicted Halley's Comet in 1304 a.d. as the Star of Bethlehem in the Scrovegni Chapel in Padua.

Vega 1 and *Vega 2* were sent by the USSR's Intercosmos agency in Moscow.

Suisei and *Sakigake* were from the Japanese Institute of Space and Astronautical Science (ISAS).

The *International Cometary Explorer (ICE)* was from NASA.

Earthbound observations by 1,000 professional astronomers and 1,200 amateur astronomers were coordinated by the International Halley Watch (IHW) in eight observing networks. IHW started collecting data in 1983 when Haley's Comet was found again. It will continue until 1989.

The chart below shows the spacecraft name, country or group of nations which launched it, launch date, date and universal time (with small Z) when it was closest to Comet Halley, distance from sun in millions of miles, distance between spacecraft and Halley's Comet in miles, and the speed in miles per hour at which the spacecraft flew by Halley's Comet.

CRAFT	Vega 1
FROM	USSR
LAUNCHED	12/15/84
FLEW BY	3/6/86 0720z
MILES TO SUN	73.47
MILES TO COMET	5523.990
SPEED	177,165.4

CRAFT	Vega 2
FROM	USSR
LAUNCHED	12/21/84
FLEW BY	3/9/86 0720z
MILES TO SUN	77.19
MILES TO COMET	4989.611
SPEED	171,796.7

CRAFT	Sakigake
FROM	Japan
LAUNCHED	1/8/85
FLEW BY	3/11/86 0418z
MILES TO SUN	79.98
MILES TO COMET	3.73 million
SPEED	168,441.3

CRAFT	Giotto
FROM	Europe
LAUNCHED	7/2/85
FLEW BY	3/14/86 0003z
MILES TO SUN	82.77
MILES TO COMET	372.8227
SPEED	153,006.4

CRAFT	Suisei
FROM	Japan
LAUNCHED	8/18/85
FLEW BY	3/8/86 1306z
MILES TO SUN	76.26
MILES TO COMET	93,827.05
SPEED	163,296.3

CRAFT	ICE
FROM	USA
LAUNCHED	8/12/78
FLEW BY	3/25/86 1031z
MILES TO SUN	99.51
MILES TO COMET	17.46 million
SPEED	145,177.2

Orbits Of Comets

A comet orbits the Sun just as planets do but the comet's path is highly elliptical while planet orbits are more nearly circular.

We only notice the comet when it is on the portion of its orbital path near the Sun and within sight of Earth. At other times it may be farther from the Sun than any known planet and out of sight.

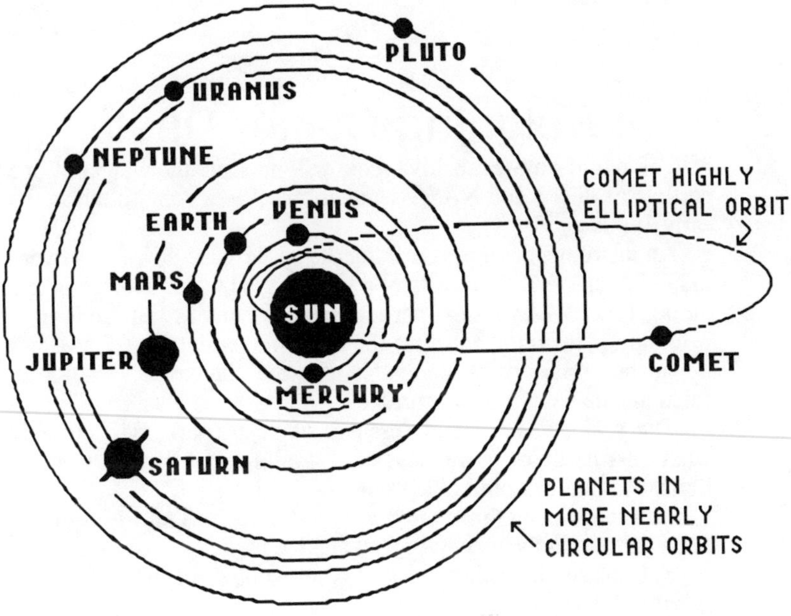

A comet probably is a lightweight clump of slush, maybe as much as 1.5 million miles in diameter, with a tiny relatively-solid nucleus surrounded by a big cloud, a nebulous coma.

A long curved plume of vapor trails the coma when the comet is close to the sun.

The whole dirty snowball probably is water, carbon dioxide, methane and ammonia.

Chunk Explodes From Comet

A chunk of rocky slush weighing 20 million pounds was blown away from Halley's Comet March 18, 1986, in a mysterious explosion.

Strong energy from the nearby Sun, rare in the comet's 76-year elliptical swing around the Sun, probably heated carbon monoxide and carbon dioxide in Halley's core. Both kinds of particles are more volatile than water and can explode if heated suddenly.

The comet was traveling away from the Sun at a distance of about 90 million miles when the blast occurred.

Although ten thousand tons of matter isn't much from a comet weighing 100 billion tons, it may have left a scar. Astronomers will be on the lookout for the mark when Halley's Comet returns to the inner area of the Solar System, near the Sun and Earth, in 2061.

Weighing Comet Dust

The world's most sensitive scale will measure the weight of one particle of dust when NASA sends it out after a comet's tail in the early 1990s.

An instrument maker is making the space scale now for delivery to NASA in 1987. It consists of a hollow tube which vibrates like a tuning fork. When a dust particle from a comet's tail lands on a weighing platform attached to the tube, the particle's weight slows down the vibrations. A computer uses the change in vibration to calculate the weight of the particle.

The scale, in the weightlessness of deep space, will measure what dust particles in a comet's tail would weigh if they were on Earth—to one-trillionth of a gram.

Astrophysicists think they will find clues to the origin of our Solar System in the weight of dust and ice particles shed by comets.

The instrument maker is tiny Rupprecht & Patashnick Co. in Albany, NY.

Earth

Australian Meteor Crater

Too bad if you were a jellyfish floating in a quiet sea near the Lake Acraman area of Australia 600 million years ago.

A passing meteorite has just smashed into your territory at a million miles per hour, sending mountain ridges of shock waves out 100 miles and throwing rocks 200 miles.

It probably bounced you around pretty good, also.

Geologists from Australia and Arizona have been poking around the area recently, looking back 300 million years before the dinosaurs. They put together the new picture, refuting arguments that the Lake Acraman crater might have formed as an erupting volcano. They found the most complete geologic record of an impact older than 70 million years.

The new data may shed light on the theory that meteorite impacts periodically throw up dust clouds which cool Earth's surface and extinguish life.

Also, if scientists can count and date all craters pockmarking Earth's face, they can estimate ages of other planets' surfaces by counting their craters.

Volcano Watching

It wasn't a rain cloud. It wasn't ozone. It was a large blob of sulpher dioxide from a Mexican volcano.

Scientists used to use conventional weather satellite cloud pictures to guess which shapes might be plumes of sulpher dioxide from volcanoes. Then they discovered the ozone-mapping spectrometer aboard the weather satellite *Nimbus 7* sees these plumes, highlighting them on maps. *Nimbus 7* has checked 16 erupting volcanoes in recent years.

Radiotelescope 1
Golf Course 0

A Delaware, Ohio, radiotelescope has a ten-year reprieve from fate as a golf course.

The radiotelescope, a meadow filled with antennas pointed toward the stars, was built by Ohio State and Ohio Wesleyan universities but Ohio Wesleyan sold the land to owners of a golf course.

Now four schools have formed the North American Astrophysical Observatory to operate the telescope. They have a ten-year lease on the land.

Earth's Lumpy Core

Where on Earth can you find mountains several miles high? Valleys six times deeper than the Grand Canyon?

No, not on the hard outer crust we call land. Not at the bottom of the oceans. Maybe we should say where *in* Earth? That lumpy blob of molten iron and nickel at the very core of our planet, is the answer.

Geologists listened to the way surface shock waves from earthquakes rattle down to the center of the Earth, bounce off the core and return to the surface. Just like a CAT scan doctors use to make pictures inside the human body, the geologists read the waves and drew pictures of the shape of Earth's core.

It had been thought the core would be a smooth sphere of red-hot metal. It actually is a rough ball.

The geologists used seismic tomography in which a computer draws a cross-section of the Earth. The technique is similar to computer-assisted tomography (CAT) scanners used by doctors to make cross-section pictures of bodies. Seismic tomography analyzes all of the shock waves from thousands of earthquakes, large and small, over ten years.

The core (*see artist's conception below*) is 4,000 miles in diameter. The Earth's crust is about 25 miles thick. The mantle is composed of viscous molten rock about 2,000 miles deep from the

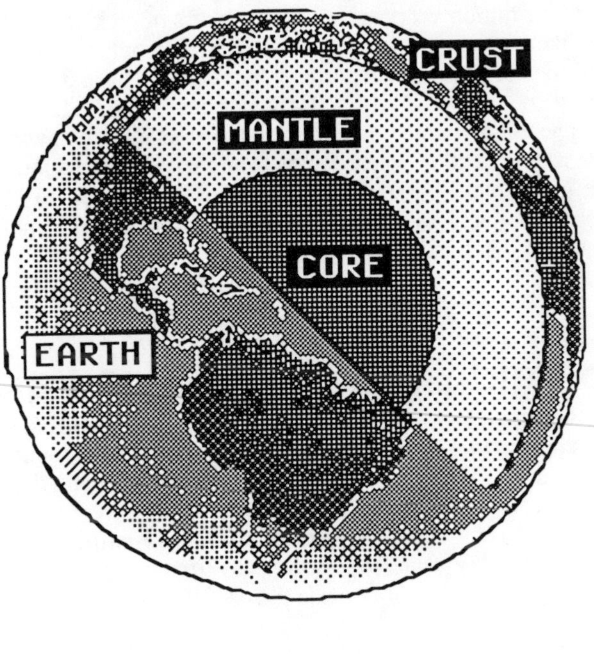

bottom of the crust to the top of the core.

The core's deep valleys are found under the parts of the crust we know as Europe, Mexico, the East Indies and the Philippine Sea. The canyon under the Philippine Sea looked to be more than six miles deep.

The mountains on the core are under what we know as central North America, the northeast Pacific Ocean near Alaska, Central America, southcentral Asia and eastern Australia. The mountain under the ocean near Alaska may be more than six miles high.

Geologists think the mountains may arise when currents flow up and down within the mantle. Hot parts of the mantle would be less dense. They would float up, away from the core like hot air from a radiator. At the top of the mantle, near the crust, some of the material may cool and sink. These up and down convection currents would pull parts of the core up into mountains and push other parts down into valleys.

Ice Balls From Space

How did we get water on Earth? Did it boil up from underground or was it brought here from space? An Iowa astrotheorist finds space-traveling slush balls brought most of Earth's water.

Scientists had thought that water boiled up from inside our planet four billion years ago as crusty layers of rock cooled. Now it looks like ice comets have been splashing into our atmosphere over the billions of years and vaporizing into clouds of water.

A glow of ultraviolet light shines from Earth's upper atmosphere. Satellite pictures show dark spots in that glowing area.

Researchers think water clouds from melted comet snow and ice block ultraviolet light and make the dark shadows 35 to 85 miles across in the pictures. Sometimes 10 spots appear per minute, stay for a few minutes and disappear.

Cosmic Dust In Blue Ice Lakes

A wealth of tiny four-billion-year-old glass spheres and enough cosmic dust to keep scientists happy for years has been found on the Greenland ice sheet by researchers from France, Denmark and the U.S.

The Earth is blanketed with more than 20-million pounds of cosmic dust every year. The shower has been going on for billions of years. You can't see it because the particles are only one-tenth of a millimeter to one millimeter across but the continuous downfall spreads a layer across the entire globe.

Some fall slowly and don't generate enough friction to melt. Others melt into little glass baubles too small to feel. Astronomers believe the unmelted dust particles are virgin samples of the gas and dust which clumped together to compose our Solar System four billion years ago.

Real original cosmic dust is hard to find. It can't be separated out from rocks and dirt on land. Some falls all the way to the ocean floor and sticks in clay where no land dirt is present. But scientists have found that an infertile source since only 10 parts of cosmic dust are found in every billion parts of ocean clay.

However, the particles fall on the Greenland ice just as they fall on everything else exposed to the sky. Dust on ice in the interior of Greenland is carried along as the glacier flows to the sea. Summer heat melts some of the glacier into blue-ice lakes in shallow basins atop the ice sheet. Cosmic dust particles collect in these lakes with a concentration of dust and glass spheres 10,000 times greater than in the ocean floor clay. The minute dust was found to be better preserved than particles in ocean clay.

Giant Fulgurite In Michigan

A giant fulgurite found in Michigan is leading some scientists to challenge the idea that dinosaurs were killed when a colossal meteorite collided with Earth 66 million years ago.

Meteorites are rich in iridium which is rare on Earth. When geologists found layers of iridium on Earth, they took that as a sign that a large meteorite carrying iridium had smashed into Earth. Supposedly the meteorite crash threw up dust, clouding the atmosphere, making life impossible for many animals and plants.

But now they have found a really big fulgurite.

A fulgurite is a small glass tube formed when lightning strikes sand, rock or oil—or when a meteorite strikes Earth. Looking through a bunch of fulgurites in southeastern Michigan, geologists uncovered the biggest yet. It's one foot in diameter and sixteen feet long.

Analyzing it, scientists found unknown minerals and odd metals. They say it appears that hits by tiny meteorites can produce unusual metals.

They speculate the rich deposits of iridium previously attributed to one big meteor might have been created by an endless string of small meteorites falling on Earth over many eons. If they are right, a new cause will have to be deduced for the extinction of dinosaurs.

Clouds From Outer Space

Those colorful wispy 50-mile-high clouds you see after sunset in the summer are water and dust from meteors vaporized as they splash into the top of our atmosphere.

The thin shining colored clouds, called noctilucent as they appear before sunrise and after sunset, can't be water vapor climbing up there from the surface of our Earth. They're just too high. We've known how high they are for the last hundred years but we haven't been able to figure how they get there.

Now a NASA physicist says the atmosphere at that altitude is cold enough for the clouds to form. Meteors pass Earth all the time. Some plunge into the atmosphere and are vaporized into water and dust.

The water and dust condense into clouds. If enough meteors splash into the warm area of the atmosphere above the equator, they can briefly cool it and permit clouds to form.

Over the polar ice caps, where temperatures are as low as minus 260 degrees Fahrenheit, it's plenty cold enough for clouds to form. Between the equator and polar ice caps, the upper atmosphere is cold enough for the clouds of meteor water and dust to condense.

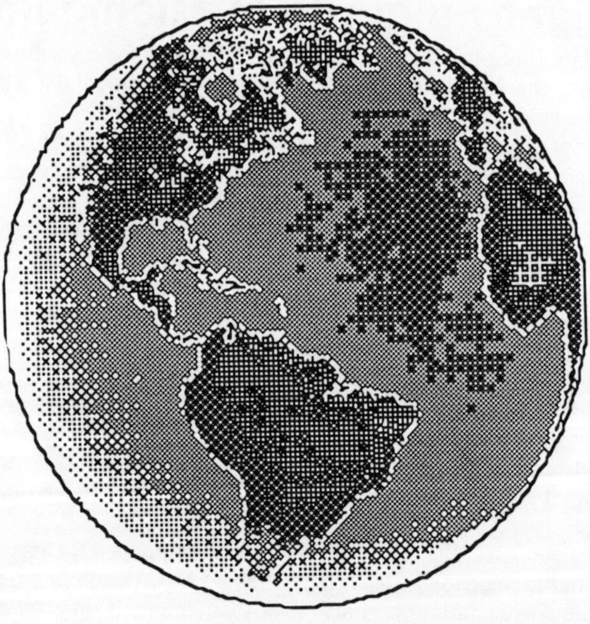

Earth Hotter Than Sun?

California researchers have found the temperature of the Earth's inner core to be 12,420 degrees Fahrenheit. The Sun's surface is only 10,400 degrees.

Of course, the Sun's center is much hotter. Scientists used to think Earth's core was around 5,000 degrees.

Meanwhile, Princeton University scientists are trying to figure out the nuclear reaction which powers stars. They have heated gas to a temperature 13 times hotter than the core of the Sun.

The gas, in a doughnut-shaped Princeton reactor, reached 360 million degrees Fahrenheit. That's about 200 million Celsius. The previous record was 80 million Celsius in 1980.

Alaska UFO

Unidentified flying objects which have been dogging a Japan Airlines cargo pilot may be unexplainable, according to the U.S. Federal Aviation Administration.

Capt. Kenju Terauchi reported November 17, 1986, while he was ferrying wine from Paris to Tokyo, something shaped like a walnut, twice as big as an aircraft carrier, and carrying a bright light followed him 400 miles in clear air over Alaska.

The 47-year-old captain, a pilot for 29 years, said the enormous object had big bulges above and below a wide brim. Two smaller UFOs flew nearby.

The captain, who lives in Anchorage with his family, was in a Boeing 747 flying more than 500 miles per hour at 35,000 feet altitude on his way from Iceland to Anchorage on a Europe to Japan flight. He had just crossed into Alaska from Canada, northeast of Fort Yukon, when he, his co-pilot and flight engineer saw three lights eight miles ahead. When Terauchi asked for permission to evade the objects, the FAA told him to drop 4,000 feet and turn around in a complete circle. The yellow, amber and green lights followed the cargo plane. FAA flight control radar reportedly showed objects as close as five miles from the airplane in view for 32 minutes.

Terauchi said the two small UFOs were only about eight feet in diameter. The larger craft had a light, but mostly was a dark globe about the diameter of two aircraft carriers placed end to end. The bigger UFO appeared on the cargo plane's weather radar screen.

Images on Air Force radar screens were labeled clutter. FAA radar images were said to be a split image of the cargo plane, a coincidence.

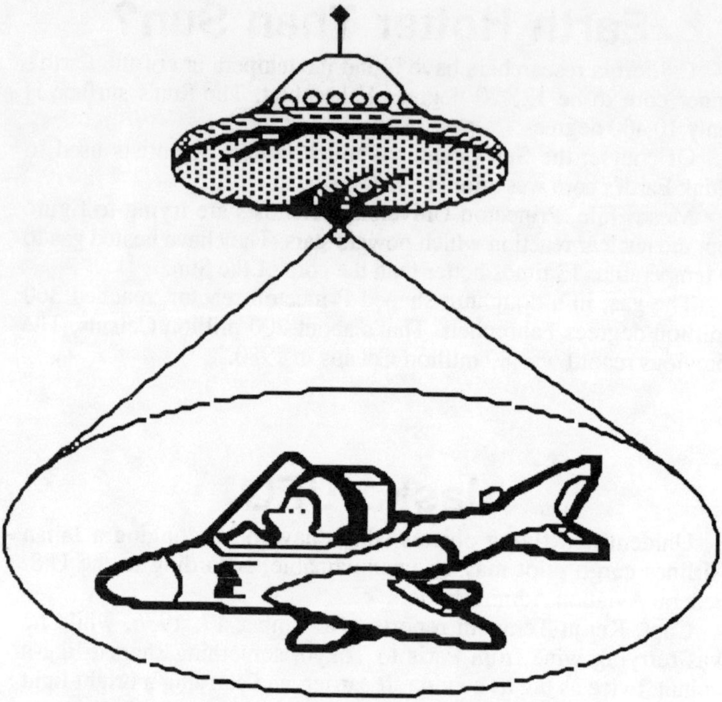

The story had just fallen out of newspapers when Terauchi was buzzed again, on January 11, 1987, by the mystery lights. FAA investigators now are speculating the lights are reflections in ice crystals catching lights from a village five miles below. The new sighting did not appear on radar screens.

But, the FAA found out it can make money out of UFO sightings.

The Anchorage office sold more than 400 packets of reports on the sighting at $194.30 each! Included was a $50 tape of the radio conversation between the Japan Air Lines (JAL) pilot and flight controllers during the 50-minute scare over the Arctic Ocean November 17, 1986, and a $56 set of glossy color pictures of radar screens.

The FAA says it can not confirm the existence of a UFO. Experts are pretty sure the radar screen revealed "split radar returns," shadows of the planes main radar echoes. Sort of like ghosts seen sometimes on a tv picture.

The FAA asked a United Airlines pilot to fly near the event. He didn't see anything.

Ancient Asteroid

A pinprick of night light became a fiery ball and, finally, executioner of the dinosaurs.

Some 65 million years ago, a five-mile-wide asteroid or comet chanced across Earth's orbit around the Sun, smashing into our planet. The impact blew vast clouds of dust and debris into the atmosphere. Sunlight was blocked from Earth's surface by the spreading cloud. Probably three-fourths of the world's animal species at that time, including dinosaurs, died. Some scientists think the scenario could have happened, but where?

A fat crater underneath Manson, Iowa, has been suggested. A big crater in southern USSR may have resulted from the crash. And now a University of Capetown geologist has found a giant bowl-shaped depression at the bottom of the Indian Ocean east of Africa.

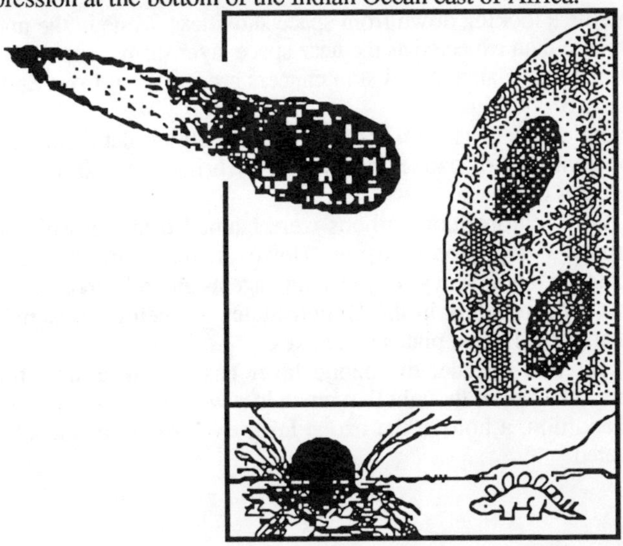

The Indian Ocean crater is known as the Amirante Basin. It is exciting scientists who believe the asteroid landed in the ocean.

Geologists say an asteroid hit because they found a thick coating of clay blanketed the earth about 65 million years ago. The clay contains an unusual concentration of iridium. That metal is rare on Earth but plentiful on meteors and asteroids.

Some think the object dunked itself in an ocean because the clay is similar to ocean crust. Others think it smashed into land because the clay contains tiny bits of quartz and feldspar. These bits are largest in the clay in western United States and Canada so some geologists think the Iowa site most likely. A few suggest more than one impact.

Meteor Splat In Blue Ridge

A big thud rattled houses, shook the ground and sounded like a giant shotgun blast underground November 10, 1986, when a meteor flashed out of the night sky and smacked into the Blue Ridge Mountains of Georgia.

Police in Tennessee, North Carolina, Alabama, as well as Georgia, reported an avalanche of phone calls after residents thought the meteor sparking through the atmosphere at dusk was a satellite crashing.

Rapid Ozone Loss

Satellites looking down from space show less ozone in the upper atmosphere than expected as the near-space layer shrinks rapidly.

Added crop damage and skin cancers are resulting from Earth's shrinking ozone filter.

Ozone protects us from the Sun's deadly ultraviolet radiation. A one percent loss of ozone from the layer brings 15,000 new skin cancer cases each year.

In 1978, chlorofluorocarbons were banned from aerosol spray cans to slow depletion of ozone. However, they still are used to make plastic foam, heavy-duty cleaning agents and refrigerants.

Fast-food retailers in the United States are being pressured to stop using foam cups, plates and boxes.

So far this decade, the ozone layer has shrunk almost three percent. It had been thought that much loss would take 40 years.

In addition, a hole in the ozone layer over Antarctica has been discovered.

Music Layer

You would need a radio to play it but you might find the music layer between Earth and space beautiful.

The magnetosphere, the farthest reach of Earth's atmosphere just before the silence of space, resounds with a symphony of sounds. Charged particles inbound from space are trapped there, creating waves. The waves shimmer with whistles and chirps which explode into a rainbow of sounds.

A Stanford University radio scientist visiting Antarctica played the magnetosphere like an instrument. He sent signals up and received a chorus of noisy response.

Milky
Way

Milky Way X-Ray Glow

Our Solar System—Sun, Earth, Moon, planets—lies off center, on one side of our Milky Way Galaxy. Thick clouds of cosmic dust between us and the heart of the galaxy prevent astronomers from seeing into the core with regular optical telescopes.

Spacelab scientists report their x-ray telescope allowed them to detect a never-before-seen unexplainable glow around the core of our galaxy.

Loren Acton was an astronaut astronomer aboard when the shuttle *Challenger* lifted *Spacelab* into space on July 29, 1985. He and six crewmates used several kinds of telescopes and other instruments to look at the Sun and deep space for eight days of the *Spacelab* 2 flight.

What was described as a "decidedly peculiar" source of x-rays was found not far from the center of the galaxy.

Whence Aluminum?

What is the Universe made of? As far as we know there are 92 elements occurring naturally. Chemistry students list these in order, by weight, starting with hydrogen as the lightest. As composition of atoms increase, weights become greater.

Astrophysicists think the Universe started with hydrogen. Nuclear reactions in common stars, such as our Sun, create more complex but still lightweight elements up to carbon and oxygen.

Heavier elements are formed in more dramatic events as in the explosion of a star.

Now a physicist in California, looking at gamma ray readings taken by an Earth-orbiting satellite, has uncovered the production of medium-weight elements such as aluminum in space beyond our Solar System near the center of our Milky Way galaxy.

The energy he spotted coming toward Earth represented aluminum-26, an isotope with a brief life of about a million years. The short life and strong energy indicated aluminum is being manufactured now. Some unusual stars or events involving stars must be producing the aluminum at the heart of our galaxy.

Milky Way Black Hole

There may be a Black Hole near the core of our Milky Way galaxy sucking in matter from the magnificent dust and gas clouds which obscure our view of the center.

The latest tantalizing info from our orbiting satellite

observatories: a bright object toward the center of the galaxy has a lot of positronium.

Under unusual conditions, electrons bind together with their opposite particles, positrons, forming positronium. Their destruction unleashes a flood of gamma rays.

The new mystery object varies in brightness of its gamma ray emissions and is huge. It seems to be more than 10,000-billion miles wide. Something that big might be a cloud around a Black Hole.

Milky Way Rifts

Anybody looking at our Milky Way galaxy—a beautiful snowfall of stars strewn in a band across a clear night sky—is struck by the noticeable breaks or rifts in the mosaic.

Those rifts in the band of stars do not mean our galaxy isn't continuous. Rather, they are vast, dark dust and gas clouds which obscure our view of thousands of stars behind them.

Astronomers using radio telescopes can see through these tiny dark nebulae to examine what's behind them.

French Messages To Space

The French government recently set up a nationwide system of computers in private homes and offices.

Called Minitel, the videotex network puts a combination of telephone and computer in every home. It's become a big consumer fad in France.

More than 2.5 million minitels already have been installed in homes and offices. The system processed 47 million calls last month!

We have videotex in the U.S. but fewer than one-half million calls were handled last month.

By dialing the nationwide computer, French users can look for jobs, make train reservations, buy groceries, test their IQs, make blind dates, question politicians, sell cars, check bank balances, play videogames, and even swap goodies through an electronic flea market.

The phone book for all of France is on the computer. A new communications medium—a videotex culture—is being formed.

One company called Cosmos Art Initiative offers a minitel service of coding messages into radio signals which are beamed toward outer space from a giant radiotelescope in central France. Thousands of messages—both funny and serious—have been sent

toward space.

Messages like "Help us understand the secret of our existence" go out every day. It makes you wonder what a recipient might think.

Lightyears

A lightyear is a measure of distance. Just as you would say your cat and dog are fifteen feet apart, or two towns are forty miles apart, objects in space—such as planets, stars, galaxies, pulsars, quasars—can be some number of lightyears apart.

One lightyear is the distance one ray of light would travel in one year in a vacuum. That's about 5.9 trillion miles or 9.5 trillion kilometers.

Space is not 100 percent vacuum but it's close enough, despite all of those beautiful clouds of dust and gas strewn through the Universe.

From one side of our own Milky Way galaxy to the other is about 100,000 lightyears.

Greek Letters Identify Stars

The bright stars on star charts are designated by Greek letters. The brightest star in a constellation is alpha or α, the next brightest is beta or β, and so on.

To identify a star, add the Greek letter to the Latin form of the constellation name. For example, the star in the constellation Orion commonly referred to as Betelgeuse is properly designated Alpha Orionis or simply α Orionis.

Here's the Greek alphabet:

α alpha	η eta	ν nu	τ tau
β beta	θ theta	ξ xi	υ upsilon
γ gamma	ι iota	ο omicron	φ phi
δ delta	κ kappa	π pi	χ chi
ε epsilon	λ lambda	ρ rho	ψ psi
ζ zeta	μ mu	σ sigma	ω omega

25 Brightest Stars

Looking up from your backyard? Here are the 25 brightest stars in the sky.

The magnitude we assign to a star depends on the amount of light it gives off and its distance from Earth. A star farther from Earth sending out a lot of light may appear brighter here than a closer star not giving off much light. A large dull star might seem weaker to us than a small bright star.

The smaller the number, the less magnitude. Thus a first-magnitude star such as Aldebaran (magnitude +0.9) seems brighter to use than Castor (magnitude +2.0).

As the numbers approach and pass zero, they become negative numbers on the magnitude scale. Thus the brightest star we see is

Sirius (magnitude –1.4), only nine lightyears away from Earth.

The very brightest objects in our sky are the Sun at magnitude –27 and the Moon at magnitude –13.

On a clear night, we can see objects as faint as magnitude 5 or 6. With binoculars we see stars as weak as magnitude 8 or 9. A six-inch telescope will show stars down to magnitude 13.

Some planets can be seen with the naked eye. Mars, for example, varies in magnitude from –2.8 to +1.6 depending upon its distance from Earth. Jupiter ranges from –2.5 to –1.4 magnitude.

Here are the 25 brightest stars in the sky in order of magnitude, including their popular names, their proper designations and their distances in lightyears from Earth, according to the Royal Astronomical Society of Canada:

MAG	NAME	DESIGNATION	DIST
–1.4	Sirius	α Canis Majoris	9
–0.7	Canopus	α Carinae	98
0.0	Alpha Centauri	α Centauri	4
–0.1	Arcturus	α Bootes	36
0.0	Vega	α Lyrae	26
+0.1	Capella	α Aurigae	45
+0.1	Rigel	β Orionis	900
+0.4	Procyon	α Canis Minoris	11
+0.5	Achernar	α Eridani	118
+0.6	Beta Centauri	β Centauri	490
+0.4	Betelgeuse	α Orionis	520
+0.8	Altair	α Aquilae	16
+0.9	Aldebaran	α Tauri	68
+0.79	Alpha Crucis	α Crucis	370
+0.9	Antares	α Scorpii	520
+0.9	Spica	α Virginis	220
+1.2	Fomalhaut	α Piscis Austrini	23
+1.2	Pollux	β Geminorum	35
+1.3	Deneb	α Cygni	1600
+1.3	Beta Crucis	β Crucis	490
+1.4	Regulus	α Leonis	84
+1.5	Adhara	ε Canis Majoris	680
+2.0	Castor	α Geminorum	45
+1.6	Shaula	λ Scorpii	310
+1.6	Bellatrix	γ Orionis	470

Deep
Space

Supernova Caught In The Act

Supernova 1987a, discovered February 24, 1987, the closest seen from Earth in 383 years, stabilized mysteriously in March 1987 but by April 1987 had started increasing rapidly in brightness again.

The supernova was expected to be brighter than any other in modern astronomy. It had the brightness of fourth-magnitude star in February 1987. It brightened to magnitude 3.63 by early April.

The smaller the number, the brighter. The North Star's is magnitude 2. Supernova 1987a was increasing by two- to three-hundredths of a magnitude per day. It now is as bright as dim stars seen with the naked eye. It may become as bright as the planet Jupiter in the sky.

Astronomers don't know when the dying star's explosion will peak or fade. Some astronomers say, if the star is big enough, they could be witnessing the birth of a black hole. But they're not sure which star exploded or how big it was.

They are learning a lot, however, about supernovae. There has not been one this close since man has had the telescope for observation. And more distant supernovae were not discovered this early in the process.

The next closest supernova was spotted in 1885 but took place about 10 times as far away as 1987a. Two much more distant supernovae, seen in 1940 and 1941, also stabilized for about 20 days before brightening further.

New Zealand amateur astronomer Albert Jones discovered the first spectacular flare-up of the supernova, known as 1987a. Professional astronomers in Chile saw the supernova start at about the same time as Jones in Nelson, New Zealand. Ian Shelton of the University of Toronto, Canada, was working at the Carnegie Institution's Las Campanas Observatory in Chile, routinely exposing photographic plates just before dawn February 24, 1987, when he noticed the supernova flare in a picture. Robert Williams at the Cerro Tololo Inter-American Observatory in Chile called the discovery the most waited-for event in the history of astronomy. Astronomers had been hoping a star would go supernova nearby ever since telescopes were invented.

What exploded?

Astronomers don't know which star blew up. At first they thought it was *Sanduleak -69º202,* a blue supergiant. Now it seems *Sanduleak* was hidden by the bright supernova. A second, smaller star near *Sanduleak* seems to be still there. Astronomers are wondering if the supernova might be a dim star they had not noticed before the explosion.

They also don't know if it is Type I, Type II, or some new third type of supernova. Type I occurs among small old stars while Type

II explosions are large young stars alone.

Astronomers don't know how a small star can release the large amount of energy needed for a Type I supernova. They suspect it happens in a binary system where two stars circle each other. One of the stars is a small dense white dwarf near the end of burning all its fuel. After pulling matter from its orbiting companion over millions of years, it accumulates sufficient mass to collapse under the weight with a great rush. The collapse compresses the core to a greater density. A new nuclear reaction starts within seconds. The star blows out a giant expanding balloon of lighter elements—a nebula—in a Type I supernova.

Large stars, on the other hand, burn fuel rapidly. Within a short lifespan of a million years, they build cores containing a lot of iron. Eventually the iron burns and energy is drained from the core. The star can't support itself against gravity. It suffers an enormous inward collapse. It becomes a neutron star as a new reaction starts. The neutron star blasts out a nebula envelope of lighter elements in a Type II supernova.

The cinder cools and fades into obscurity unless enough mass remains for the core to become a black hole. A black hole is an extraordinarily dense concentration of matter with gravitational pull so strong not even light can escape.

Supernova 1987a may be growing brighter because it is expanding in size more quickly than it is cooling off. The bubble of high-energy matter is flowing away from the star at about 6,000 miles per second. That envelope now is some 50 billion miles in diameter.

If there had been a planet like Earth orbiting the dying star, it long since would have been incinerated.

The envelope's temperature has cooled from 36,000 degrees Fahrenheit to 7,200 degrees. Astronomers know about the cooling because the supernova changed from blue to red. It was blue at its hottest and red as it cools.

If the bubble stays hot it will continue to expand. If cooling becomes predominant, it will shrink and start to fade from our view. Astronomers say the supernova may continue to brighten or it could start fading immediately.

It will fade eventually

A supernova is a star that suddenly increases by billions of times in brightness. After days or weeks it slowly fades away. Supernovas occurred in our Milky Way galaxy in the year 1054, in 1572, and in 1604. In those, a star became bright enough to be seen in daylight and then faded months later. The famous *Crab Nebula* is the remnant of material spewed into space by the 1054 supernova.

Lots of supernova remnants can be seen in deep space today with telescopes. Supernovae found since invention of the telescope have occurred in galaxies millions of lightyears from Earth.

A supernova radiates more energy in a few days than our Sun does in 100 million years. Usually, as in the *Crab Nebula* supernova, the burned-out cinder left behind after the explosion is a neutron star only a few miles in diameter. It is enormously dense and consists mainly of neutrons, but not the same as a black hole. A pulsar is a pulsating neutron star.

Cosmologists calculate the Universe came into existence 15 billion years ago when the single-atom element we know as hydrogen was created. Since then, nuclear fusion in stars has changed some of the original hydrogen into helium and heavier elements.

Supernovas have produced heavy elements and pushed them out to make new stars and planets. A supernova may even have exploded shortly before the formation of our Solar System. Elements spewed out from such an explosion could have become part of the Sun, Earth, and living things we know.

Large Magellanic Cloud

Supernova 1987a is in the Large Magellanic Cloud, a neighboring galaxy near our own Milky Way. The Large Magellanic Cloud is 163,000 lightyears away. Light seen today from the supernova left there 163,000 years ago traveling at 186,200 miles per second.

The Milky Way galaxy, a flat spiral of stars adrift in space, is about 100,000 lightyears across. The Large Magellanic Cloud, then, is more than 60,000 lightyears farther away, beyond the distant edge of our own galaxy. The Large Magellanic Cloud can only be seen from Earth's Southern Hemisphere.

Stars exist for millions, even billions, of years. They form when minute dust particles and gas floating in space chance to clump together, gathering sufficient weight to crush themselves into a nuclear furnace. Lightweight simple atoms of hydrogen and helium at the core of the star provide fuel. The thermonuclear reactions generate light and other energy and create new heavier atoms. The ancient star which exploded into Supernova 1987a is one among billions in the Large Magellanic Cloud.

After burning the dust and gas at the core steadily, for millions of years, fuel at the heart of the star is used up. It starts to collapse in on itself. The outer layers of the star, including newly created heavier elements, start to fall inward. The crush leads to a bright burst of explosion.

For several weeks the star will burn very brightly, consuming fuel from its outer layers. It will give off as much light as an entire galaxy.

Life cycle

This supernova explosion symbolizes the life and death cycle of the Universe. The star dies. In its wake is a burned out cinder which, if small enough, will glow warmly for eons. Or, if it's large enough, will collect new fuel and become a black hole.

Meanwhile, the heavy elements in the outer layers of the dying star, blown off by the supernova explosion, drift away to seed the Universe with new matter. As some of this matter clumps together later, it becomes new solar systems.

Supernova 1987a is believed to have a mass more than eight times the Sun. Scientists don't know whether so much will be blown off in the nebula that the core will be too small to produce a black hole. In that case, the core will become a neutron star, a dense mass of neutrons a few miles in diameter.

A dozen new supernovae are spotted by astronomers in far-away galaxies across the Universe each year. A galaxy like the Milky Way may experience a new supernova every 15 to 50 years. The last one seen in the Milky Way, however, was Kepler's Supernova in 1604, five years before Galileo gave astronomy the telescope. Johannes Kepler saw it with his naked eye.

A total of four have been recorded in our galaxy in the last 1,000 years. The nearest supernova seen since Kepler's was in 1885 and ten times farther away than Supernova 1987a in the Large Magellanic Cloud.

Small Magellanic Cloud

Both small galaxies close by our Milky Way—the Large Magellanic Cloud and the Small Magellanic Cloud—can be seen from the Southern Hemisphere by naked eye. Ferdinand Magellan found them in 1519.

The Large Magellanic Cloud is 39,000 lightyears in diameter, less than half as wide as the Milky Way. It is in the area of sky we refer to as the Dorado Constellation.

The Small Magellanic Cloud is under 20,000 lightyears in diameter. It is farther away from us at 196,000 lightyears distance in the Toucan Constellation.

Astronomers suspect the Small Magellanic Cloud may be two galaxies lined up. If so, the Mini Magellanic Cloud would be some 33,000 lightyears beyond the Small Magellanic Cloud in line from Earth. A lightyear, roughly six trillion miles or nine trillion kilometers, is the distance light travels in a year.

The Large Magellanic Cloud may have been closer to the Milky Way, drifting past a billion years ago. A trail of gas and dust which astronomers call the Magellanic Stream extends across our sky. It probably is matter ripped away from the Large Magellanic Cloud by Milky Way gravity.

Scientists aren't sure whether the Clouds are satellite galaxies of our own galaxy, forever in orbit around the Milky Way, or simply passing by on a one-time chance encounter on their way through the Universe.

Detectors at Mt. Blanc in Europe, in a saltmine at Fairfax, Ohio, and at Kamioka, Japan, reported neutrinos from the supernova, the first ever from outside our Solar System.

NASA watch

NASA's Deeps Space Network station near Canberra, Australia, started watching radio waves from the supernova. Early readings clocked particles flying out of the supernova at 6,200 miles per second. That's about 1/30th the speed of light.

A 110-ft. antenna at Tidbinbilla near Canberra and a 210-ft. antenna at Australia's Parkes Radio Observatory 200 miles away have been tied together by microwave link. The two form one giant radio telescope in a process called interferometry.

Radio astronomy, which receives microwave radio signals generated by natural objects in space rather than visible light, can distinguish objects hundreds of times better than optical telescopes.

The last time the two antennas in Australia were linked was during the Voyager 2 visit to the planet Uranus in 1986. NASA Deep Space Network also has stations near Madrid, Spain, and Goldstone, California.

IUE satellite

A telescope aboard a nine-year-old satellite orbiting Earth has been looking at ultraviolt radiation from the supernova.

An 18-inch telescope aboard the *International Ultraviolet Explorer (IUE)* satellite—the largest telescope now operating in space—was focused on the Large Magellanic Cloud for a look at Supernova 1987a.

IUE revealed the star *Sanduleak* did not explode. Visible light had looked like *Sanduleak* was gone but *IUE* still could see *Sanduleak* in ultraviolet light. *IUE* also could see the previously-known near neighbor of *Sanduleak*. So, scientists are wondering what third star exploded.

NASA was able to move quickly to study the Supernova 1987a because it keeps plans for contingencies on hand. U.S. scientists, as well as researchers in other countries, are polled for so-called targets of opportunity. Future plans include looking at exploding stars, comets and other unusual objects.

Supernova 1987a is giving off very bright ultraviolet light. Ultraviolet rays are a form of light with shorter wavelengths and greater energy than visible light. Ultraviolet rays are absorbed by the Earth's atmosphere so they can't be studied from telescopes on the ground.

Even though Supernova 1987a is in a neighboring galaxy, IUE may reveal useful information about our Milky Way galaxy. The Milky Way and the Large Magellanic Cloud each may have a galactic corona, a hot cloud of gas and dust at the outer fringes of the galaxy. These regions of space have not been explored well and the ultraviolet light from the supernova may shed some light on these coronae.

Supernova 1987a is like a bright lightbulb turned on outside the corona of our galaxy. Astronomers will learn something about the corona from how much ultraviolet light is absorbed as it passes through.

IUE, operated jointly by the European Space Agency, the United Kingdom Science and Engineering and Research Council, and NASA, controlled from the Goddard Space Flight Center, Greenbelt, MD, still works well despite being years past its design life and without some of its original gyros.

Picturing A Supernova

Scientists used to think that an exploding star would throw off its gases in a smooth bubble expanding outward. But new pictures of a supernova seem to show bumps and holes in the shell of gas.

Images of the supernova Cassiopeia A were made in 1981, 1983 and 1985 using the Very Large Area (VLA) radio telescope in New Mexico. VLA is the world's largest permanent radio telescope with 27 dish antennas, each 82 feet in diameter, near Socorro, New Mexico.

Not satisfied with the first results, astronomers passed the radiotelescope data through a Hollywood supercomputer which, most of the time, animates special effects for movies and rock videos.

The Cassiopeia A supernova gas bubble is huge—10,000 times the width of our Solar System. The computer enhanced images showed the shell to be pockmarked with bumps and holes where smaller bubbles of heavier gases poked it from inside.

Naked Supernova

Three unusual exploding stars have been spotted in as many years.

The uncommon supernovae are giving astronomers their first look into the hearts of dying stars.

Late in a life of billions of years, a star usually explodes in a quick bright flash of light. The outer ball of hydrogen gas surrounding the star's core blows out. It spreads in all directions from the star, preventing us from seeing what may be happening at the core.

We can study the core only sometime later when the gas shell has dissipated. First-hand viewing of crucial events in the death of the star are left to imagination.

Now, fortuitously, we find the naked supernova. The gas shell has been stripped away before the final ball of fire. Astrophysicists think this may happen when two stars are orbiting each other. The companion's gravity sucks the hydrogen away before the final core collapse.

Without the mask of hydrogen gas, the natural light waves and radio waves from the dying core can travel unblocked to Earth. If we are looking at the right spot in the sky, we see an intimate view of the details of the aging star in the early time of its cataclysmic explosion.

A naked supernova has been seen in each of the past three years.

Christmas Supernova

An exploding star, 300 million lightyears away in the distant galaxy NGC 1667, is being called a Christmas supernova by scientists who found it Dec. 22, 1986.

Colin Okada, an undergraduate astronomy student at the University of California, first spotted the supernova in the constellation Eridanus during a nightly sky sweep.

The university astronomers examine 400 galaxies by computer-controlled telescope each night, searching for previously-unseen space objects. In six years of looking, four supernovae have been found.

A supernova is the final violent end of a large star. With hydrogen fuel running out, its center collapses. Helium and other remaining atoms under that pressure fuse together, forming heavier elements. Nuclear and gravity energy are released. The star blows itself apart.

Backyard astronomers need a small home telescope to see the Christmas supernova. It's directly overhead at midnight but too faint for the naked eye. The constellation Eridanus is south of the better-known Taurus constellation.

The supernova is at the center of the NGC 1667 galaxy of stars.

Fastest Double-Star System

It takes only 11.3 minutes for one star to whip around another in the fastest double-star system found so far.

Compare that with the month it takes Earth's Moon to go around once. Or the year it takes Earth to circle our star, the Sun.

In the newly-discovered binary, or double-star, system, the outer star is a *white dwarf*. It's only about 80,000 miles from the central companion, a *neutron star*. That's only one-third the distance from Earth to the Moon.

The neutron star is tiny—only about five miles in diameter—but it is massive. It weighs a third more than our Sun which is about 865 thousand miles in diameter. Even our own planet Earth is nearly 8 thousand miles in diameter. The neutron star's extraordinary mass gives it even greater gravity pull than the Sun.

The white dwarf orbiting the neutron star is about the size of Earth. Its mass is less than one-tenth that of our Sun.

Changing theories. Previously, astronomers would have expected the neutron star to have captured a passing white dwarf and tugged it inward. The white dwarf would have spiraled down quickly until it was destroyed by the neutron star. However, this pair doesn't seem to work that way.

It looks like this white dwarf was captured millions of years ago, while it still was a normal star like our Sun. In theory, the normal star was floating by when gravity from the neutron star caused tides in its surface gases and, finally, a bulge. The normal star lost momentum and moved only a short distance away from the neutron star. It was pulled into a highly-elliptical orbit around the neutron star. With each pass around the neutron star, tidal bulges were drawn out and the star settled into a circular orbit around the neutron star. At that point, it might have taken ten hours for the normal star to circle the neutron star.

Matter was sucked from the surface of the star leaving a white-dwarf core. Over time, the entire outer atmosphere of the star was blown away, leaving only a naked white dwarf star. After this loss of mass, it began to spiral closer to the neutron star. Each orbit around the central star became faster.

The European *Exosat* satellite spotted the speed demons.

Binary stars. Two stars circling each other are called *binary* stars. Such double star systems are very common in our galaxy and across the Universe. A single star alone, such as our Sun, is more unusual than systems with two or more stars orbiting each other. Some astronomers suspect our Sun may have a binary companion somewhere nearby. They are searching for a deep-space object so small it will take very strong binoculars to see it—if it exists.

Eleven minutes is by far the shortest known orbital period. The

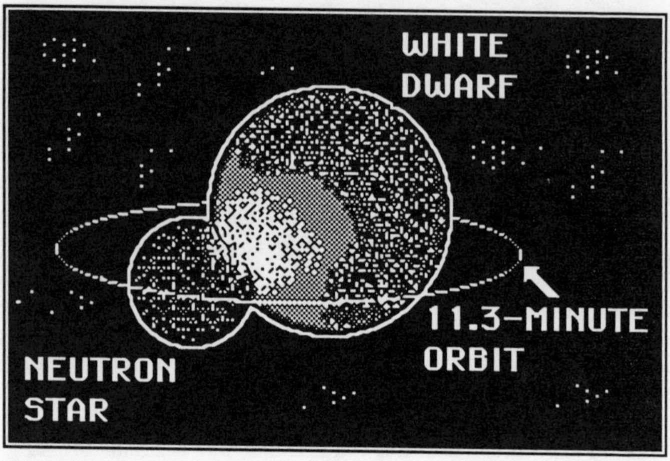

two distant stars forming the binary system are at the heart of a very bright source of x-rays far, far away—some 20,000 lightyears away—in the constellation of *Sagittarius*.

The bright source has been named *x1820-30* after its position in the sky. One of the two stars composing x1820-30 is the dying white dwarf gripped firmly by the neutron star.

Neutron star. A neutron star is a very dense cinder of a large star which exploded into supernova and collapsed.

Our *Milky Way* galaxy has about 50 x-ray sources similar to x1820-30. They are powered by gas falling from one star onto a neutron star. The gas falls quickly, at about one-third the speed of light. This releases an enormous amount of energy from the neutron star—about 10,000 times the output of our Sun.

The neutron star in x1820-30 is a stellar corpse, probably five miles across but not more than thirteen miles across. By comparison, Earth is 7,926 miles across.

As the gas from the white dwarf races onto the neutron star, it becomes very hot. X-rays are produced. To the *Exosat* satellite, which sees x-rays, it looks very bright.

Exosat discovered that the x-rays from x1820-30 send out a pulse or blip every eleven minutes. This indicates the companion star blocks *Exosat's* view of the neutron star as the companion swings around every eleven minutes.

White dwarf. This orbital period is so short that the companion star must be tiny—for a star. It probably is only about the size of Earth but not more than three times bigger than the Earth. That's somewhere between 8,000 and 24,000 miles in diameter. It is a white dwarf made of pure helium.

These two stars—the neutron star and the white dwarf—would fit easily between the Earth and the Moon. Yet, the total mass of the system is greater than our Sun. A lot of weight in a very small space!

This remarkable double star lives at the center of a *globular cluster* called *NGC 6624*. A globular cluster is a dense conglomeration of very old stars which have accumulated near each other over the eons. Stars are far apart in our part of our Milky Way galaxy so there's not much chance of two stars colliding around here anytime soon. But, in a globular cluster, stars are a million times closer together.

The binary system x1820-30 probably was formed when the neutron star attracted a normal star called a *red giant*. A red giant is a normal star, similar to our Sun, but much later in its life span. Our Sun is not yet a red giant.

The European x-ray observatory satellite *Exosat* was launched aboard a Delta rocket in May 1983.

Meanwhile, another binary star system has been found recently. In it, a normal star takes eight to nine hours to orbit a neutron star. It may be similar to the 11-minute binary but at an earlier point in its life.

Supernova Seen

Astrophysicists have seen a star start to die. A nondescript star 60 million lightyears from us in the galaxy known as *M99* began its final moment of glory on May 17, 1986. Buried in the Virgo cluster of galaxies, the star exploded into a supernova.

Supernovas blossom unexpectedly, last a few weeks and fade as a star dies. Those brief moments are important. That's when the heavier chemical elements are created. The supernova death of an old star gives birth to new stars and planets as the chemical elements created within the star are blasted into space.

Scientists have come across remnants of supernovas before but were frustrated at not seeing one from beginning to end. Berkeley, California, physicist Frank Crawford, looking over pictures shot May 17, comparing them with photos made May 9, spotted the bright new light from the *M99* supernova. He is on a search team using a 30-inch telescope near Berkeley to photograph 1,000 galaxies every night

A lightyear is the distance light travels in one year, about six trillion miles.

Since Crawford's discovery, the extraordinarily close supernova *1987a*—discovered February 24, 1987—has given astronomers another chance to see a star start to die.

Four Stars Locked In Orbit

An Arizona scientist has discovered one of astronomy's rarest phenomena, four stars locked in orbit around each other.

The quadruple system, about one quadrillion miles from Earth in the constellation Leo, is one giant gravity web. The ensnared four are pulling and tugging at each other, producing magnificent tidal effects, spewing out spectacular flares from hot gas eruptions on their surfaces.

The system is two pairs of stars circling each other. At the heart, two stars actually touch each other. One is larger than the other making them look like a lopsided dumbbell. They rotate every six hours around a common center of gravity.

Circling them is another pair about a billion miles away. The stars in the second pair are about 1.4 million miles apart. They rotate every 18 hours around a common center of gravity. The outer pair circles the inner pair every 20 years.

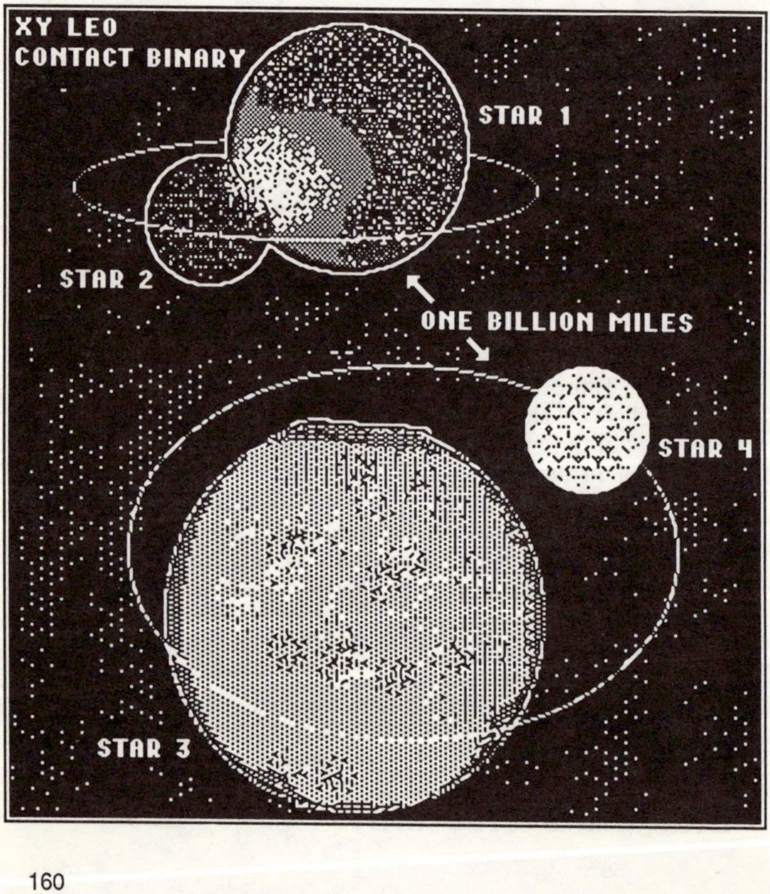

The four may have formed from one disk of hot gas and dust billions of years ago. But, because the orbit planes of the two pairs are slightly different, the two sets may have formed in separate clouds of gas and dust. Some powerful gravity source could have torn a pair from its original position, moving it near the other pair, leaving all four locked in a new gravity embrace.

There are billions of stars in the Universe, most in pairs called binary stars. About a dozen quadruples have been found. Our Sun is unusual in being a single star with no companion.

Binary Pulsar

There are seemingly endless numbers of stars strewn through the Universe. Most appear to be in pairs, companions circling each other.

A neutron star is a very tiny dense cinder remaining after a large star exploded into supernova and collapsed. Many astrophysicists think pulsars are neutron stars.

Pulsars appear to send out high-energy bursts, thus the name pulsar. These blips come and go at regular intervals. Some so-called millisecond pulsars blink rapidly, sending out many bursts of energy each second.

A binary pulsar is a pulsar and a companion star locked in orbit around each other.

One such binary pulsar is known as PSR1855+09. Part of its system is a millisecond pulsar turning on and off 200 times per second. Its companion, a white dwarf star, is a large lightweight sphere of hot helium.

Pulsar Clock

A pulsar 12,000 lightyears away may replace Earth-bound atomic clocks as our most reliable timepiece.

The pulsar, known as *1937+21* for its location on astronomers' sky maps, spins on its axis at what seems to be a very steady 642 twirls a second.

That's unusual. Pulsars, dense cinders left when stars burn out, usually spin faster and faster as time passes.

When a star, at least half again as large as our Sun, reaches a stage late in its life, it explodes into a supernova. That blows off its outer layers of gas. The remaining core collapses under its own weight, squeezing down to a ball as small as five miles in diameter.

Like an ice skater who spins faster by pulling in legs and arms, the compressing of the core makes it spin faster. The atoms in the core are squeezed so tightly they break down and cease to exist as atoms. All that remains are neutrons packed so densely a thimbleful would weigh a billion tons. Thus, a neutron star is born. A neutron star radiates large quantities of energy in the form of x-rays, radio waves, light waves.

A pulsar is formed when the powerful gravity of a neutron star captures a second star and holds it in orbit around the neutron star. As that second star circles the neutron star, it momentarily blocks our view of the radio waves coming from the neutron star. From our observation post at Earth, the neutron star's radio waves appear to blink on and off regularly. This pulsing on and off led to the name pulsar.

The pulsar *1937+21* was discovered in 1982. It has not seemed to vary in spin rate since then. That makes it the most stable thing we know of anywhere.

With no detectable fluctuation in its spin rate, the pulsar is more regular than atomic clocks built on Earth. The official U.S. atomic clocks are kept in a National Bureau of Standards laboratory at Boulder, CO. They measure the natural vibrations in cesium atoms to mark time. However, they do fluctuate. For instance, two atomic clocks vibrating at nine billion cycles per second have been found to deviate by a few billionths of a second after only a few years. The pulsar *1937+21*, on the other hand, hasn't deviated since 1982.

By the way, if the original star had been large enough, the collapse to a neutron star would have continued. The gravity then would become so powerful even light could not escape it. From that point on, it would be a *black hole*.

Quasar. Pulsar. But, Noisar?

What are you going to call a noisy neutron star when household words like quasar and pulsar already are taken? *Noisar*.

At least that's what the Los Alamos National Laboratory in New Mexico is calling *Sco X-1*. That deep space object is one of the most familiar x-ray sources in our Milky Way galaxy. In fact, it was the first x-ray source ever recognized. Now we see it has one of the most unusually noisy signals coming from space.

Energy radiated from pulsars, quasars and neutron stars usually has a natural rhythm to it. Noisars, on the other hand, also are including a lot of seemingly random incoherent energy in what they send out.

Astrophysicists want to make sense of it. Until they do, the new class is called noisars to distinguish them from plain-jane rhythmic neutron stars.

Among the 100 billion stars in our galaxy, a mere few hundred x-ray sources have been spotted. And, of those, only a tiny handful are noisars.

Like other big signals from outer space, *Sco X-1* is a binary system—two stars circling each other. One is a cinder of a burned out ancient star—a neutron star.

The neutron star's companion throws helium and hydrogen into space where it is sucked in by the neutron star's extraordinary gravity. The blast when the gas crashes into the neutron star at high speed is such a large explosion that tremendous energy flies out into space, including x-rays.

Feeding Quasar Black Holes

Vast amounts of gas and dust are concentrated near the open maw of a hungry black hole, fueling a new quasar when two spiral galaxies collide.

Birth explosions of ten quasars in the far reaches of the Universe are being studied by California Institute of Technology astrophysicists.

A black hole is a small dead star so extraordinarily dense that its powerful gravity won't even let light escape. It sucks in loose gas and dust in space and even strips matter from stars which come too close as they float through space. Black holes may be the generators powering quasars.

3,600 quasars. Quasars are extremely bright deep-space objects. Some 3,600 have been seen since the first was identified in 1963. Most are among the objects farthest away from Earth seen so far. That makes them the oldest since light we see today would have taken longer to travel the greater distance to Earth. The most powerful quasars are the brightest objects we see in the Universe.

The core of a quasar is as large as our entire Solar System. It's as if our star, the Sun, ballooned out to a diameter large enough to encompass all of the known planets. A quasar core gives off ultraviolet light as well as visible light. One quasar is brighter than an entire galaxy of billions of stars.

Ultraluminous galaxies. Astronomers decided some time ago that quasars are formed when galaxies collide. A billion suns smashing into and becoming enmeshed with a billion other suns forms a new very-luminous infrared star system. These ultraluminous infrared galaxies are the intermediate step as a quasar is formed.

The infrared galaxy gives off lots of energy in the infrared part of the spectrum but little visible light. In fact, after the collision, the new galaxy gives off ten times more heat than light. The new infrared galaxy has a distorted central disk of stars as the two sets merge. Some stars are stripped off each of the parent galaxies and are thrown out into long tails streaming away from the new galaxy.

Tremendous energy. A great deal of gas and dust is left in the center of the new galaxy. This fuels the quasar being born. When such unusually large quantities of dust and gas fall onto a black hole, tremendous energy is generated at the edge of the black hole producing the intense light we see from a quasar.

A small amount of visible light can be seen coming from the intermediate-stage infrared galaxy. That visible light looks more like light from a quasar than light from a single spiral galaxy or a galaxy merely forming new stars.

We see more infrared from the intermediate-stage galaxy than visible light because the dust and gas filling the new galaxy block our view of the core. The dust and gas absorb and are heated by the ultraviolet and visible light being given off by the quasar forming at the core. Eventually most dust and gas is blown away and a mature quasar can be seen giving off visible light.

The ten infrared galaxies studied were photographed by the Infrared Astronomical Satellite which orbited Earth in 1983. The ten galaxies range in distance from 200 million to one billion light years from Earth. That's 1.2 billion trillion to 5.9 billion trillion miles from our planet.

Hearing Strings On A Radio

Cosmic strings may sound like Brazilian beachwear but they're far from it. Space theorists figure they are leftovers from some Universe-wide cataclysms in past eons and that we should be able to see them.

Cosmologists try to decide how the Universe came to be. However it came to be, some think it has undergone large-scale shifts in *phase* from time to time. These phase shifts may have altered, somehow, space and time.

Cosmologists are on the lookout for strange objects in deep space, millions or billions of lightyears from Earth. Those objects, if they exist, are called *strings*. They would be remnants of previous space and time, sort of a ragged edge where some previous incarnation of our Universe didn't mate up exactly with the present version. These relics of an unknown past might be moving through our Universe at near the speed of light. They would shed a great deal of light on how our Universe came to be if one or more could be found.

Strings would have so much energy in our Universe they probably would give off radio signals. Looking at the skies with radio telescopes, cosmologists hope to see radiowave-emitting objects shaped like long, thin strings.

Now some cosmologist can be heard saying, "Ah ha!" They are looking at the deep space radio source known as G357.7-0.1. It appears to be string-like and some are saying it might be a cosmic string. Next they have to calculate the rate at which its filaments are moving through space to see if they are near light speed.

Missing Gravity Source

Astronomers say they now can account for more dust, gas and gravity in space than before but there's still something missing.

They have calculated for some time that there is a lot more gravity in the Universe than could be coming from the stars. They used to think loose gas and dust generated the extra gravity.

Now they think the Universe has been more efficient than they expected in making stars and galaxies out of dust and gas.

There still are huge intergalactic clouds of dust and gas but new studies have seemed to indicate that 90 percent of known matter already has been gathered into stars. The 10 percent left over in clouds of dust and gas is not enough to account for the calculated total gravity of the University.

Astronomers say there must be an unknown source of gravity. They will look for it other than in the gas and dust clouds.

Cosmic Diamonds

Trillions of tiny diamonds—older than Earth itself—have been found in a meteorite which plopped into a Mexican cornfield in 1969.

A Chicago researcher says the diamonds, so small they're almost diamond dust, were created in the fiery supernova explosion of a dying star long before our Solar System even existed. He says if scientists can figure out exactly how the diamonds formed, manufacturers may be able to make better industrial diamonds.

Put 20,000 of these diamonds in a row and you would have the width of a human hair. They are at least 4.5 billion years old.

Three other meteorites packed with the microscopic gems have been found on Earth. Diamonds are carbon with atoms arranged in a unique crystal lattice. Diamond dust found previously in meteorites was thought to have formed in the shock of impact with Earth.

Carbon might condense from expanding and cooling gases in the upper atmosphere of a dying red-giant star. Stellar wind, an outpouring of matter from the star, could have created a fountain of diamonds blowing out into the cosmos. When the star finally exploded to its death, the carbon dust could have been compressed into diamonds and embedded in matter strewn across space.

Researchers didn't expect to find diamonds in the Mexican meteorite. They washed away layers of black carbon, turning it white. The white layer turned out, surprisingly, to be carbon—diamonds. The honeycomb interiors of the diamonds hold unusual forms of xenon and krypton which flagged the meteorite as not from our Solar System. The diamonds are tiny time capsules, preserving chemical samples from the environment of the dying star.

Blue Arcs: Largest Structures

Colossal bands of new stars splashed across distant clusters of galaxies in mysterious glowing blue arcs found by astronomers are the largest structures ever seen in the Universe.

Three of the curving lines of stars, in patterns never seen before, have been spotted by Stanford University astronomers at the U.S. National Optical Astronomy Observatories at Kitt Peak, Arizona.

Very big. The blue arcs are like ropes of stars, each about 1.9 million trillion miles long. By comparison, Earth and other planets circle one star we know as the Sun. Our Sun is only one among a billion or so stars composing our Milky Way galaxy. The newly-found blue arcs are three times longer than the diameter of the Milky Way galaxy.

A cluster of galaxies might hold more than a billion billion stars. The blue arcs are so long they curve through three clusters of galaxies.

Very far away. The blue arcs are 19 billion trillion miles from Earth, looping through galaxy clusters called Abell 370, Abell 2218 and 2244-02.

Visible light from the arcs was first noticed from Kitt Peak Observatory 10 years ago. But the light was too faint for study. New more-sensitive telescope analysis shows the arcs curved like ropes around certain bright elliptical galaxies within the clusters of galaxies. Astronomers think the bright elliptical galaxies grow by consuming smaller galaxies passing by.

Stars gathered in a field shaped like these long blue arcs have never been seen before. They are peculiar to astronomers, giving headaches to theorists trying to understand what caused the arcs.

Very mysterious. The visible blue light means the arcs probably are composed of new stars which could have been formed recently by an especially-powerful black hole passing through the galaxy clusters. A black hole is a dead star so small and dense that its powerful gravity won't even let light escape.

If not formed by a passing black hole, the stars may have been created when a blast wave of gas, dust and energy from an unknown gigantic cataclysm disturbed the galaxy clusters.

If an entire galaxy exploded, the outbound shockwave might create a band of stars along its leading edge. But astronomers say there wasn't enough gas and dust in the area to compose stars as bright as they see.

One of the arcs appears broken into clumps and knots which have been noticed before in new-star systems.

To some astronomers, the arcs look like the jets of gas squirted across space by quasars. Those unusually bright objects send out strong radio signals with the gas jets. The trouble is, quasar jets usually don't give off as much visible light as these arcs. And the arcs don't seem to have the radio energy of a quasar.

Could an arc be a galaxy pulled and stretched out by gravity from a bigger galaxy passing by? Probably not, since the three clusters of galaxies in the vicinity don't have enough gravity on hand to do the job.

Nor do the arcs match the predicted size or shape of cosmic strings which some astronomers hope to see some day.

Whatever these long curving ropes of stars are, they are the largest visible structures yet seen from Earth.

Who's Calling, Please?

Is there intelligent life Out There sending signals we can detect? Every ten days or so, alarms sound as out-of-the-ordinary radio waves from deep space are received by sky sweeping receivers in Massachusetts. So far they have been false alarms.

A dish-shaped antenna, 84 feet in diameter, points toward the sky. Every two minutes, Earth's rotation causes the antenna to point toward a different part of the sky. Radio signals are received around the clock and analyzed by a bank of 128 computers.

The computers analyze the two-minute chunks of radio waves from space by dividing the spectrum of energy received into 8.4 million separate frequencies. Each frequency is analyzed to see if it seems to contain artificial signals. Is there something in the signal not attributable to background radiation caused by natural phenomena in the sky?

The antenna, receivers and computers are part of the Harvard-Smithsonian Observatory. Once a day an operator changes the north-south elevation of the antenna just enough for it to look at different two-minute chunks of sky.

At one time or another during the year, nearly 80 percent of the entire sky passes in view of Harvard, Massachusetts. Over seven months, the antenna sees it all.

The computers stumble across something which might be intelligent every ten days on average. Human observers always examine such data but, so far, nothing intelligent has passed their tests. And the sweep goes on.

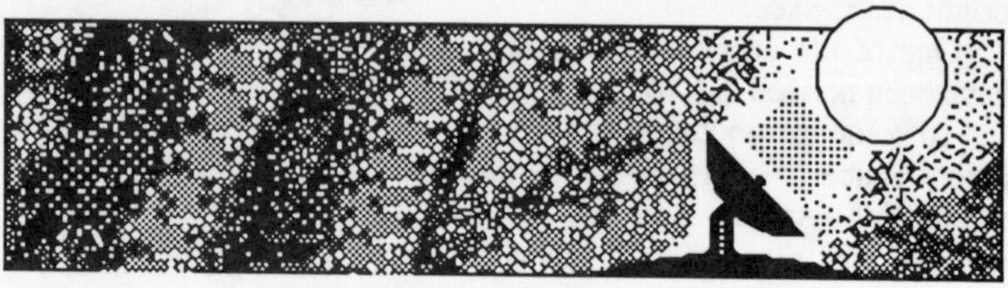

Deep Space Telescopes

A new generation of observatories in space, above the sight-fogging atmosphere of Earth, will open previously unseen views of the Universe in the 1990s.

The four high-tech instruments with strange-sounding acronyms will be Gamma Ray Observatory (GRO), Advanced X-Ray Astrophysics Facility (AXAF), Space Infrared Telescope Facility (SIRTF) and the already-famous Hubble Space Telescope (HST).

Man has studied the heavens since ancient times but the view has always been made fuzzy by the ocean of air between Earth's hard surface and the near-vacuum of space. Only just now are we able to put entire astronomical observatories in space where the seeing is not blurred by an atmosphere. Now we know that objects in space—stars, galaxies, quasars, pulsars—give off radiowave energy, beyond visible light.

The electromagnetic spectrum of radio waves extends from very low frequencies to extremely high. Our ability to tune in those radio waves has grown in recent years. It's only been in this century we have used any radio, starting at the longwave end of the spectrum. Today's AM broadcast stations transmit signals in the mediumwave portion of the spectrum. FM stations use VHF transmitters and TV stations use VHF and UHF. Cooking ovens are microwave and police speedtraps are radar transmitters and receivers, all part of the energy spectrum. Visible light we receive with our eyes is flanked in the spectrum by infrared and ultraviolet light we can't see. We do, however, have the capability to capture infrared and ultraviolet energy on photographic film. The same for x-rays.

Only recently have we been able to make radio receivers, covering the UHF to gamma ray part of the spectrum, small enough to send into space. Here's what we can expect from NASA with current technology:

▲ The *Gamma Ray Observatory* will explore the strongest part of the energy spectrum. It will receive a much wider range of wavelengths than earlier observatories. GRO will return clues to the question of whether most gamma rays arriving at Earth from across

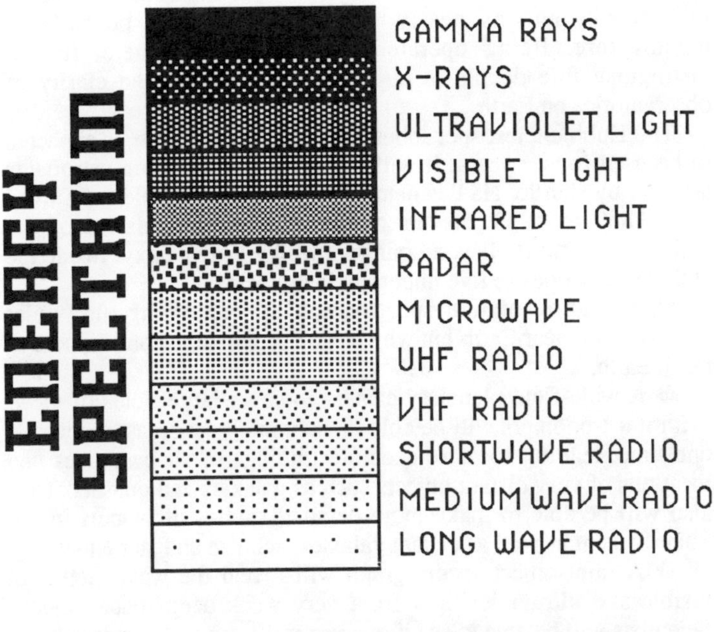

the Universe come from quasars and pulsars. Or are there other objects in our universe sending out gamma rays.

▲ The *Advanced X-Ray Astrophysics Facility* will see the x-ray part of the energy spectrum. It will be 1,000 times more sensitive than earlier instruments. AXAF will be an international observatory used to study the x-rays found to be coming from almost every known object in the Universe. It will help us look at planets and common stars as well as exotic objects such as neutron stars, black holes, quasars and the hearts of active galaxies. We already have seen a quasar 12.4 billion lightyears away. AXAF will let us see even more distant quasars.

▲ The *Space Infrared Telescope Facility* receiver will span the infrared part of the spectrum. It too will be 1,000 times more sensitive than earlier equipment. So far, we only suspect the existence of planets around stars not far from our own Sun. SIRTF may be the telescope that shows us planets around other stars. If we see planets in star systems outside of our Solar System, the search for extraterrestrial intelligence, known as SETI, surely will become more clearly focused.

▲ The *Hubble Space Telescope* will be the largest telescope ever sent to space. Looking at visible light and ultraviolet light, it will expand the observable volume of the Universe by several hundred times.

The primary mirror in HST is 2.4 meters or about 94.5 inches in diameter and as nearly perfect as presently possible to manufacture. Being operated above the atmosphere, it will distinguish fine details in objects with ten times the clarity of observatories on Earth.

The Hubble telescope, about 370 miles above Earth, is expected to be used for 20 years. It will be serviced by visiting astronauts arriving by shuttle. HST is named for Edwin P. Hubble, who lived from 1889 to 1953. He discovered that the boundaries of the Universe extend far beyond our own Milky Way galaxy. Inside the Hubble telescope are five important instruments:

● A high-resolution spectrograph for examining ultraviolet light arriving near Earth but which can't penetrate the atmosphere to reach Earth.

● A wide-field planetary camera to cover large fields of view. With it astronomers will be able to examine large areas of space at one time, leading to more accurate findings of shapes, sizes and locations of very distant objects such as galaxies and quasars. They also will be able to make more detailed studies of planets in our Solar System as well as distant galaxies, nebulae and star clusters.

● A faint-object spectrograph will reveal the wavelengths of visible and ultraviolet light from very weak deep-space objects. Scientists will be able to see if quasars really are at the core of some galaxies. Or how comets are changed by approaching the Sun.

● A high-speed photometer will measure light from objects in space. It will see fluctuations in brightness down to blinks lasting only ten millionths of a second.

● A faint-object camera, developed by the European Space Agency, will make pictures of faint galaxies at unprecedented distances from Earth. It also will see closer objects in greater detail.

The operations control center for the Hubble telescope will be at NASA's Goddard Space Flight Center at Greenbelt, MD. Science data from the Hubble telescope will be sent to the Space Telescope Science Institute at Johns Hopkins University in Baltimore, MD. NASA's Marshall Space Flight Center at Huntsville, AL, will be in charge of maintenance of the telescope.

Photograph The Night Sky

The sky is most mysterious, most intriguing at night.

You can look up from our position at the bottom of an ocean of air anytime. But daytime, revealing little of space beyond the Sun and Moon, doesn't hold the mystique for most. Nighttime, when the heavens twinkle, captures our fancy.

We look with the naked eye. We see more with binoculars. We view the most through telescopes. At some point we want to record the excitement of what we are seeing. We want to make photos.

Astrophotography. Men have been collecting images of light from space for 100 years. Capturing starlight for our picture archives is a popular part of the hobby of astronomy. Here's how to go about it:

✓ Any *camera* may catch some light coming from space but your best bet will be one which makes negatives or slides of 35mm or larger size. The 35mm single-lens reflex (SLR) is most popular. A larger-format camera will permit bigger enlargements.

✓ You'll need to be able to mount your camera on a *tripod* to steady it. A shaking or vibrating camera will make ugly pictures. The camera should have a "bulb" or B setting for time exposures and a way to manually trigger the shutter. Some exposures require the shutter to be open for hours so beware of a battery-powered shutter. Batteries do run down.

✓ The *lens* on your camera depends on what you want to shoot. Besides the "normal" 50mm lens which came with your camera you may want to consider a wide-angle 28mm or 35mm lens. Telephoto lenses of 128mm and 200mm also will be useful.

✓ The *film* you use depends on what you wish to photograph and the format of your camera's negative. Brands to consider include Kodak Kodachrome, Ektachrome, Tri-X, Pan-X, Technical Pan and VR; Fuji Fujichrome; 3M Scotch; and Agfa XRS and RS.

More light is received from nearby planets in our Solar System so you can use the luxury of slower films which have a finer grain.

Stars, galaxies, clusters and other objects outside our Solar System are so far away that precious little light arrives at Earth. You either need extreme lengths of exposure time or a fast film. You will opt for the faster film but at the expense of pictures with more and bigger grain. Some hobbyists *hypersensitize* film to make it more sensitive.

Use color slide film rather than color negative film for color shots.

Kodak Tri-X 400 film is a good negative film for black-and-white photography. The grain is not fine but it sees weak light well. A good fine-grain black-and-white negative film is Kodak 2415 Technical Pan. It has an ISO rating (ASA rating) of 100. If there's enough light, Kodak 2415 gives fine detail and it can be hypersensitized.

The length of time the shutter is open depends on the type of film you use. Faster films, with higher ISO numbers, will record more light in a short time. One drawback comes when you discover that faster films also fog more quickly from background sky light. Slower films can be have longer exposures but they, too, eventually will fog.

Your exposure time will depend upon how bright the sky is at your location. Shoot a series of pictures to find the right exposure. If you are using a 400-speed film and a 50mm lens try exposures of 15, 20, 25 and 30 minutes for your first experimental shots. If you use 1000-speed film with a 50mm lens open the shutter for 5, 6, 7, 8, 9 and 10 minutes. Films slower than 400 speed may be exposed for longer than 30 minutes.

Stars can look like nothing more than white specks on a frame of film. The processor may not be able to align the first frame for cutting and mounting. Either ask to receive the developed roll of film uncut and unmounted, or shoot a normal well-lit scene for the first frame on a roll to be used as a placement guide.

A *telescope* often is needed to record details of space objects. Prime focus, eyepiece projection and afocal photography are common ways to use a camera and telescope in astrophotography.

In prime focus the telescope is used to guide the camera and to make the photograph. The camera is attached to the telescope and you look through an eyepiece to make sure the telescope is tracking the object you want to record. Exposures of deep-sky objects often run from 30 minutes to three hours. During that time you must watch the eyepiece to make sure the object doesn't drift out of view.

Eyepiece projection is used to magnify objects in Sun, Moon and planet photos.

In afocal photography the camera sits on its tripod and is pointed into the telescope's eyepiece. Since you are adding the magnification of the camera's lens to the magnification of the eyepiece, overall magnification is greater than in simple eyepiece projection.

What to shoot. You can make pictures of the Moon, stars, meteors, comets, planets, galaxies, nebulae, deep sky objects, constellations. If it can be seen it can be photographed. Sometimes it can be photographed even if you can't see it otherwise:

✓ Meteors aren't hard to record. Mount your camera on a tripod. Set the shutter for time exposures on "bulb" or B. Leave the shutter open for a period of time ranging from four minutes up to many hours depending upon the ISO (formerly called ASA) rating of the film you use. If you are using an ISO 400 film, the shutter-open period should range from four minutes to one hour.

Such long exposures will make light sources in the night sky look like streaks. The streaks or meteor trails will be straight near the North Star.

Meteor showers, which happen several times a year, are best for seeing because there are more rocks per hour.

✓ Stars, seen in photographs as trails or streaks, are recorded in the same manner as meteors. To go beyond pictures of trails across the sky requires a *clock drive* to match your camera to the rotation of the Earth. Objects in the night sky move slowly from east to west as our globe turns. A telescope with a clock drive moves in a direction opposite to this star motion. The telescope's clock drive is used by attaching your camera to the telescope.

✓ Constellations are recorded by pointing the camera toward the appropriate part of the sky and making exposures of several seconds each. Check your star map. If the constellation you want is above your horizon, point your tripod-mounted camera toward it. If you use a "normal" lens of about 50mm or a "wide-angle" lens of from 20 to 50mm, use fast film. Adjust the f-stop or aperture control either one or two stops below the lens' fastest setting. For example, if your lens opens up to f-1.4, set it at 1.8 or slower for the exposure. Using the B shutter setting and a cable shutter trigger, make a series of six exposures starting at 10 seconds and working up to one minute. You will have exposures of 10, 20, 30, 40, 50 and 60 seconds. Exposures longer than 60 seconds with a normal lens will cause the stars to look like streaks. With a wide-angle lens, limit exposures to under two minutes. A normal 50mm lens should be used to record Orion, the constellation with the beautiful Barnards Loop, Orion and Horsehead nebulae.

✓ Our Milky Way galaxy and its gigantic dust clouds are best photographed through a 28mm wide-angle lens.

✓ The planets of our Solar System are bright in the sky but look like star points of light because they are small and so far away. The position of a planet against the background of stars changes rapidly and can be photographed with a brief exposure 50mm lens. If you want to make detailed pictures of planets you will need a telescope to magnify the small object.

✓ Comets are similar to planets when it comes to photography except that their long orbital paths only bring them in close enough for us to see them for short periods from time to time.

✓ Asteroids are tiny chunks of rock in orbit around the Sun like planets. Since they are only a few miles across they are quite faint and only the brightest ever become visible to the eye. Like planets, they look like star points of light and they appear to move against the background of stars. A telescope must be used to make photographs of asteroids.

✓ Variable stars can be seen to regularly change in brightness over time. Some can be seen by eye but they should be photographed using a telescope. Try to shoot one picture when the star's brightness is near maximum and another near minimum. To start, take pictures of the same area ten days to three weeks apart.

✓ Binary stars are most common in the sky. These double stars actually are two or more orbiting each other. The stars in one binary are often different colors but you need a telescope to separate the single star point of light into two or more stars.

✓ Star clusters, beyond our Solar System and even beyond our galaxy, are deep sky objects requiring a telescope to make photographs.

✓ Galaxies, beyond our own Milky Way, must be photographed through a telescope to obtain detail.

✓ Pictures of the Moon are best during its first or third quarters. Contrast is best then so details look better. A full moon can be the worst time to shoot. Use a telescope to record craters.

✓ The Sun is far and away the brightest object in our sky and, therefore, the most dangerous for photographers. Never look directly at the Sun with the naked eye or through an unfiltered camera or telescope. Filters for sunlight include white-light filters which cut the amount of light by 99.9 percent and show the Sun as it looks to the naked eye. And hydrogen alpha filters which reveal only hydrogen alpha light from the Sun. White-light filters show the grainy surface of the Sun and sunspots. Hydrogen alpha filters allow you to see solar prominences and flares. A telescope is needed to shoot details of the Sun's face.

Which Lens To Use...

What *is the deep-sky object you may wish to photograph.* When *is the season on Earth.* Where *refers to photographer's position in the Northern Hemisphere or Southern Hemisphere of Earth. Lenses are 28mm, 50mm, 135mm and 200mm.*

WHAT	WHEN	WHERE	LENS
Big Dipper	Spring	North	28, 50
Andromeda galaxy	Fall	North	50, 135, 200
Sagittarius	Summer	North	28, 50
Orion	Winter	North	50
Star clouds	Summer	South	28, 50
Pleiades	Winter	North	135, 200
Scorpius	Summer	North	50
Coma star cluster	Spring	North	135
North American nebula	Summer	North	50, 135
Hyades	Winter	North	135
Summer Triangle	Summer	North	28

California nebula	Winter	North	135
Large Magellanic Cloud	Summer	South	135
Small Magellanic Cloud	Summer	South	135, 200
Rosette nebula	Winter	North	135
Eta Carina Nebula	Summer	South	135
Hyades & Pleiades	Winter	North	50

Which Film To Use...

Good films for astrophotography. Not all films listed fit all cameras. First column shows brand name and type. Second column shows sky object to photograph.

Kodak	
Kodachrome 25	planets
Kodachrome 64	planets
Ektachrome 64	planets
Ektachrome 100	planets
Ektachrome 200	planets/deep space
Ektachrome 400	planets/deep space
Tri-X 400	planets/deep space
Pan-X 100	planets
Technical Pan 100	planets
103a-E	deep space
103a-O	deep space
103a-F	deep space
VR 100	planets
VR 200	planets/deep space
VR 400	deep space
VR 1000	deep space
Fuji	
Fujichrome 100	planets
Fujichrome 200	planets/deep space
Fujichrome 400	planets/deep space
Fujichrome 400D	planets/deep space
Fujichrome 1600D	deep space
3M	
Scotch 640	deep space
Scotch 1000	deep space
Agfa	
XRS 100	planets
XRS 200	planets/deep space
XRS 400	deep space
XRS 1000	deep space
100 RS	planets
400 RS	deep space
1000 RS	deep space

The Astronomy Revolution

So many advances in space research have spilled out of observatories and laboratories during this last half of the 20th Century that astronomy has undergone a revolution. Many astronomers think of this as a golden age.

Here's what's been happening:

Optical astronomy: Traditional astronomers are awash in new techniques of faint-object detection including an exciting range of electronic-imaging gear, computers, video screens and more-sensitive photographic emulsions.

Using today's standard telescope, an astronomer can see fainter, more distant objects than ever before. And he no longer is limited to looking only for visible starlight. Now his instruments let him study deep space in new parts of the energy spectrum, including infrared and ultraviolet light.

Infrared astronomy. Scientists have known about infrared radiation from deep-space objects since the 18th Century. In the early 1960's, they got down to serious infrared astronomy research.

Sensitive new detectors gave astronomers eyes to explore the infrared-light portion of the electromagnetic spectrum to better understand how stars are born, how young stars evolve, and the vast dense clouds of dust surrounding stellar maternity wards.

Radio astronomy. Radiotelescopes routinely study natural signals from stars, planets, the dust and gas between stars in our own galaxy, and sources beyond our Milky Way galaxy.

Radio astronomers have mapped the galaxy and discovered quasars, pulsars, and a large number of complex organic molecules in interstellar space.

Radar astronomy has been used to measure and map Venus and Mercury in our own Solar System.

Ultraviolet, x-ray, gamma-ray astronomy. Earth's atmosphere is a murky opaque barrier to most infrared and ultraviolet light, gamma rays and x-rays. So satellites orbiting above the ocean of air—as well as high-altitude balloons and one-shot rockets—have become popular platforms from which astronomers observe those exotic wavelengths.

The U.S., USSR and others have launched several observatories to orbit since 1962 to get a better look at ultraviolet light and x-rays coming from deep-space objects. Now we know a lot more about such strange discoveries as hot young stars and black holes.

Space shuttles ferrying telescopes to orbit, men in space stations, the Hubble Telescope, all will bring more advances.

Space
Futures

SPACE FUTURES

WHEN THE EXCITEMENT CONTINUES

1987

USA

•NASA will attempt to launch six satellites

USSR

•USSR may launch 90-100 satellites

•USSR will add two large modules to the *Mir* second-generation space station in Earth orbit

•USSR will test fly a huge new heavy-lifter rocket

•USSR amateur operators may send two *RS* radio satellites to Earth orbit

•USSR will launch *Aktivny-1K* Earth satellite to study magnetosphere and the *Magion* satellite will be popped into separate orbit by *Aktivny-1K*

•USSR will orbit *Biosat* carrying two monkeys, several rats, other small animals to study effects of weightlessness

•USSR will launch *Gamma-1* USSR-French telescope to study deep-space gamma rays and x-rays

•USSR will send the *Roentgen* telescope module up to be added to the *Mir* space station complex

•USSR will launch an Earth-resources module to be attached to *Mir*

Europe

•European Space Agency may launch seven of its previously-grounded *Ariane* rockets

•Europe will decide to go ahead with construction of *Hermes*, a European space shuttle

Japan

•Japan will launch spacecraft to Earth orbit using its *H-1* rocket

1988

USA

•U.S. shuttle returns to space with three to five flights scheduled between June and December

•The U.S. *Tracking and Data Relay Satellite TDRS-C* will be carried to space in *Discovery*, the first shuttle to fly following the 1986 *Challenger* explosion. It will be shuttle flight number 26

•A secret military satellite will be carried aloft on shuttle flight number 27

•A CIA spy photo satellite will ride in shuttle flight number 28

•Another radio relay satellite, *TDRS-D*, will be ferried to space in shuttle flight 29

•The Hubble Space Telescope may be carried to orbit late in 1988 on shuttle flight 30

•A U.S. funeral home will use a private Space Services Inc. *Conestoga* rocket to launch to Earth orbit a space mausoleum holding 10,000 tubes of ashes
•American Rocket Co., a private launch firm, may send its *Industrial Launch Vehicle* on a test flight to space from California
•Utah college students will see their *NUSAT 2* amateur satellite carried to Earth orbit in a shuttle
•U.S. radio amateurs may see their Phase 3-B *OSCAR* satellite ride along on a European *Ariane* rocket

USSR

•USSR will send spacecraft to orbit Mars with cameras and other instruments
•USSR will send spacecraft to orbit Mars' moon Phobos with cameras and other instruments
•USSR will test fly a new space shuttle to Earth orbit strapped to its huge new heavy-lifter rocket
•USSR may launch 90-100 Earth satellites
•USSR will add modules to the *Mir* space station
•USSR will loft the *Granat* French-USSR telescope to space to study x-ray stars
•A new *Astron* type of spacecraft will go into general use transporting equipment to Mir
•USSR will send up the first two of four *Interbol* French-Russian craft

Europe

•European Space Agency will fire eight *Ariane* non-reusable rockets to orbits
•Europe will move ahead on construction of the *Hermes* European space shuttle
•Europe will go forward on construction of *Columbus*, the European space station

Japan

•Japanese amateur radio operators will launch *Japan Amateur Satellite 2*
•Japan will launch other spacecraft to orbit using its *H-1* rocket
•Japan will plan construction of a small space shuttle
•Japan will go forward on ground construction of its module for the U.S./Canadian/Japanese/European international space station to start space construction in 1994

China

•China uses *Long March* rocket to launch private U.S. satellites
•China may launch its first manned spaceflight

India

•India may attempt another flight using the *Augmented Satellite Launch Vehicle*

1989

USA

•U.S. spaceprobe *Voyager 2* will fly within 3,000 miles of the planet Neptune

•U.S. space shuttle is scheduled for ten flights

•A group of ultraviolet telescopes, *ASTRO-1,* will remain with the shuttle during flight number 31

•A secret military satellite will be sent to stationary orbit from flight 32

•The *Magellan* spacecraft to study Venus will be carried aloft in shuttle flight 33

•A secret military Star Wars space-weapons *Spacelab* will be in flight 34

•Flight 35 will carry a materials-science research payload plus two military navigation *Navstar* satellites

•Another secret military satellite will be sent to stationary orbit from flight 36

•The *Galileo* spacecraft, carried to Earth orbit in a shuttle in October or November 1989, will be sent to drop a probe into the atmosphere of Jupiter

•The U.S. Air Force may start using separate expendable *Delta 2* rockets to send up a fleet of *Navstar* satellites

USSR

•USSR may launch 90-100 satellites

•USSR will add modules to the *Mir* space station

•USSR will send up the last two of four *Interbol* French-Russian craft

•Two *Progress* and two *Magion* craft will study the magnetosphere

Europe

•European Space Agency will fire eight *Ariane* non-reusable rockets to orbits

•Europe will move ahead on construction of the *Hermes* European space shuttle

•Europe will go forward on construction of *Columbus,* the European space station

Japan

•Japan will launch scientific and other spacecraft to Earth orbit using its *H-1* rocket

•Japan will continue construction of a small space shuttle

•Japan will go forward on ground construction of its module for the U.S./Canadian/Japanese/European international space station to start space construction in 1994

China

•China will launch communications satellites for Iran and Sweden

•China uses *Long March* rocket to launch private satellites

•China may launch manned spaceflight

India

•India will fly the *Augmented Satellite Launch Vehicle*

1990

USA

•U.S. space shuttle is scheduled for eleven launches in 1990

•The *Ulysses* spacecraft, carried to Earth orbit in a shuttle, will fly close to the Sun to study our star

•The British *Skynet 4* military communications satellite will be ferried to orbit in a U.S. shuttle
•India's weather and communications satellite will be carried to space in a U.S. shuttle

USSR

•USSR may launch 90-100 satellites
•USSR will add modules to the *Mir* space station
•USSR will send up an *Aelita* Russian-French craft to examine very-short-wavelength radio signals from deep space
•The *Rascas* radiotelescope with 100-ft. antenna may be launched
•A *Radio Astron* satellite launched between 1990-1992 will use a 33-ft. antenna to receive signals from deep space

Europe

•European Space Agency will carry satellites to Earth in *Ariane* non-reusable rockets
•Europe will move ahead on construction of the *Hermes* European space shuttle
•Europe will go forward on construction of *Columbus,* the European space station

Japan

•Japan's *Muses-A* spacecraft will fly by the Moon several times, dropping off a small lunar orbiter on the first pass
•Japan will launch scientific and other spacecraft to Earth orbit using its *H-1* rocket
•Japan will continue construction of a small space shuttle
•Japan will go forward on ground construction of its module for the U.S./Canadian/Japanese/European international space station to start space construction in 1994

China

•Canada will send a satellite to orbit from China
•China uses *Long March* rocket to launch private satellites
•China may launch manned spaceflight

India

•India will fly the *Augmented Satellite Launch Vehicle*

Australia

•Australia may open its commercial spaceport at Weipa in the state of Queensland for launches by other countries and private flights

1991

Worldwide

•Sunspot activity peaks

USA

•U.S. space shuttles will make ten flights to orbit

USSR

•USSR may launch 90-100 satellites
•USSR will add modules to the *Mir* space station

Europe

•French-U.S. satellite to map Earth's oceans
•Europe will test *Hermes,* a European space shuttle

•European Space Agency will carry satellites to Earth in *Ariane* non-reusable rockets

•Europe will go forward on construction of *Columbus*, the European space station

Japan

•Japanese-U.S. satellite will study the Sun

•Japan will launch scientific and other spacecraft to Earth orbit using its *H-1* rocket

•Japan will continue construction of a small space shuttle

•Japan will go forward on ground construction of its module for the U.S./Canadian/Japanese/European international space station to start space construction in 1994

China

•China uses *Long March* rocket to launch private satellites

•China may launch manned spaceflight

India

•India will fly a more powerful *Augmented Satellite Launch Vehicle*

1992

USA

•U.S. *Mars Observer* will orbit the planet with cameras and other instruments

•The *Mariner Mark II* spacecraft, leaving Earth with cameras and instruments to fly beyond Mars to the Asteroid Belt for pictures of asteroids, also will fly alongside a comet for three years

•U.S. space shuttles will make ten flights to orbit

USSR

•USSR will send Mars spacecraft including small Mars Rover vehicle, atmospheric balloons, ground penetrators, cameras, soil analysis lab, map maker

•USSR may launch 90-100 Earth satellites

•USSR will add modules to the *Mir* space station

Europe

•Europe will fly its *Hermes* space shuttle

•European Space Agency will carry satellites to Earth in *Ariane* non-reusable rockets

•Europe will go forward on construction of *Columbus*, the European space station

Japan

•Japan starts using its new large *H-2* non-reusable rocket to ferry satellites to Earth orbit

•Japan will continue construction of a small space shuttle while planning for a larger shuttlecraft

•Japan will go forward on ground construction of its module for the U.S./Canadian/Japanese/European international space station to start space construction in 1994

China

•China uses *Long March* rocket to launch private satellites

•China may launch manned spaceflight

India

•India will fly its *Augmented Satellite Launch Vehicle*

1993

USA

•U.S. will launch three deep-space telescopes to Earth orbit sometime in the 1990s: the *Gamma Ray Observatory*, the *Advanced X-Ray Astrophysics Facility* and the *Space Infrared Telescope Facility*
•U.S. space shuttles will make ten flights to orbit

USSR

•USSR may launch 90-100 satellites
•USSR will add modules to the *Mir* space station

Europe

•Europe will fly its *Hermes* space shuttle
•European Space Agency will ferry satellites to Earth orbit with *Ariane* rockets
•Europe will go forward on construction of *Columbus*, the European space station

Japan

•Japan will launch scientific and other heavy spacecraft to Earth orbit using its *H-2* rocket
•Japan will continue construction of a small space shuttle and planning for a larger shuttle
•Japan will go forward on ground construction of its module for the U.S./Canadian/Japanese/European international space station to start space construction in 1994

China

•China uses *Long March* rocket to launch private satellites
•China may launch manned spaceflight

India

•India will send spacecraft to orbit using its *Augmented Satellite Launch Vehicle*

1994

USA

•U.S. space shuttles will make ten flights to orbit
•NASA plans to start carrying space station construction materials to space

USSR

•USSR will send Mars spacecraft with larger Rover and instruments
•USSR may launch 90-100 Earth satellites
•USSR will add modules to the *Mir* space station

Europe

•Europe may send to Earth orbit its *Columbus* space station module to be part of the U.S./Canadian/Japanese/European international space station
•Europe will fly its *Hermes* space shuttle

•European Space Agency will ferry satellites to Earth orbit with *Ariane* rockets

Japan

•Japan will launch scientific and other spacecraft to Earth orbit using its *H-2* rocket

•Japan will continue construction of a small space shuttle and planning for a larger shuttle

•Japan will finish construction of its module for the U.S./Canadian/Japanese/European international space station

China

•China uses *Long March* rocket to launch private satellites

•China may launch manned spaceflight

India

•India will send spacecraft to orbit using its *Augmented Satellite Launch Vehicle*

1995

USA

•U.S. will start building a first-generation space station in Earth orbit

•U.S. space shuttles will make ten flights to orbit

USSR

•USSR may launch 90-100 satellites

•USSR will add modules to the *Mir* space station

Europe

•Work will continue in space on the *Columbus* space station module, part of the U.S./Canadian/Japanese/European international space station

•Europe will fly its *Hermes* space shuttle

•European Space Agency will ferry satellites to Earth orbit with *Ariane* rockets

Japan

•Japan may fly its small space shuttle to Earth orbit attached to its large *H-2* non-reusable rocket

•Japan will start construction of a large space shuttle

•Japan also uses its large *H-2* rocket to ferry satellites to Earth orbit

•Japan may launch its module for the international space station

China

•China uses *Long March* rocket to launch private satellites

•China may launch manned spaceflight

India

•India will send spacecraft to orbit using its *Augmented Satellite Launch Vehicle*

1996

USA

•U.S. space shuttles will make ten flights to orbit

USSR

•USSR will send Mars spacecraft with larger Rover, instruments, and ability to pick up rock and soil samples and return them to Earth

•USSR may launch 90-100 Earth satellites
•USSR will add modules to the *Mir* space station

Europe

•Work continues in the *Columbus* space station module, part of the U.S./Canadian/Japanese/European international space station
•Europe will fly its *Hermes* space shuttle
•European Space Agency will ferry satellites to Earth orbit with *Ariane* rockets

Japan

•Japan will launch spacecraft to Earth orbit using its new space shuttle and its powerful *H-2* rocket
•Japan may work in its module at the international space station
•Work will continue on the large space shuttle

China

•China uses *Long March* rocket to launch private satellites
•China may launch manned spaceflight

India

•India will send spacecraft to orbit using its *Augmented Satellite Launch Vehicle*

1997

Worldwide

•Sunspot activity drops to eleven-year minimum

USA

•U.S. space shuttles will make ten flights to orbit

Europe

•Work continues in the *Columbus* space station module, part of the U.S./Canadian/Japanese/European international space station
•Europe will fly its *Hermes* space shuttle
•European Space Agency will ferry satellites to Earth orbit with *Ariane* rockets

USSR

•USSR may launch 90-100 satellites
•USSR will add modules to the *Mir* space station

Japan

•Japan will launch spacecraft to Earth orbit using its space shuttle and powerful *H-2* rocket
•Japan may work in its module at the international space station
•Work will continue on the large space shuttle

China

•China uses *Long March* rocket to launch private satellites
•China may launch manned spaceflight

India

•India will send spacecraft to orbit using its *Augmented Satellite Launch Vehicle*

1998

USA

•U.S. space shuttles will make ten flights to orbit

•U.S. may launch a robot spacecraft to Mars to return in 2001 with rock samples

USSR

•USSR may send spacecraft beyond Mars to the Asteroid Belt to land and return samples of an asteroid to Earth

•USSR may launch 90-100 Earth satellites

•USSR will add modules to the *Mir* space station

Europe

•Work continues in the *Columbus* space station module, part of the U.S./Canadian/Japanese/European international space station

•Europe will fly its *Hermes* space shuttle

•European Space Agency will ferry satellites to Earth orbit with *Ariane* rockets

Japan

•Japan will launch spacecraft to Earth orbit using its space shuttle and its powerful *H-2* rocket

•Japan may work in its module at the international space station

•Work will continue on the large space shuttle

China

•China uses *Long March* rocket to launch private satellites

•China may launch manned spaceflight

India

•India will send spacecraft to orbit using its *Augmented Satellite Launch Vehicle*

1999

USA

•U.S. space shuttles will make ten flights to orbit

•U.S. may finish building its first-generation space station in Earth orbit

USSR

•USSR may launch 90-100 satellites

•USSR will add modules to the *Mir* space station

Europe

•Work continues in the *Columbus* space station module, part of the U.S./Canadian/Japanese/European international space station

•Europe will fly its *Hermes* space shuttle

•European Space Agency will ferry satellites to Earth orbit with *Ariane* rockets

Japan

•Japan will launch spacecraft to Earth orbit using its space shuttle and its powerful H-2 rocket

•Japan may work in its module at the international space station

•Work will continue on the large space shuttle

China

•China uses *Long March* rocket to launch private satellites

•China may launch manned spaceflight

India

•India will send spacecraft to orbit using its *Augmented Satellite Launch Vehicle*

2000

USA

•NASA will launch space shuttle flights as well as several Earth satellites

USSR

•USSR may launch 90-100 satellites

•USSR will add modules to the *Mir* space station

Europe

•Work continues in the*Columbus* space station module, part of the U.S./Canadian/Japanese/European international space station

•Europe will fly its *Hermes* space shuttle

•European Space Agency will ferry satellites to Earth orbit with *Ariane* rockets

Japan

•Japan may fly its larger space shuttle to Earth orbit attached to a large non-reusable rocket

•Japan also will launch spacecraft to Earth orbit using the H-2 rocket and the smaller space shuttle

•Japan will work in its module at the international space station

China

•China uses *Long March* rocket to launch private satellites

•China may launch manned spaceflight

India

•India will send spacecraft to orbit using its *Augmented Satellite Launch Vehicle*

Index

$19.95
320 pages
11"x8-1/2"
paperback
perfectbound
ISBN 0-86668-063-2

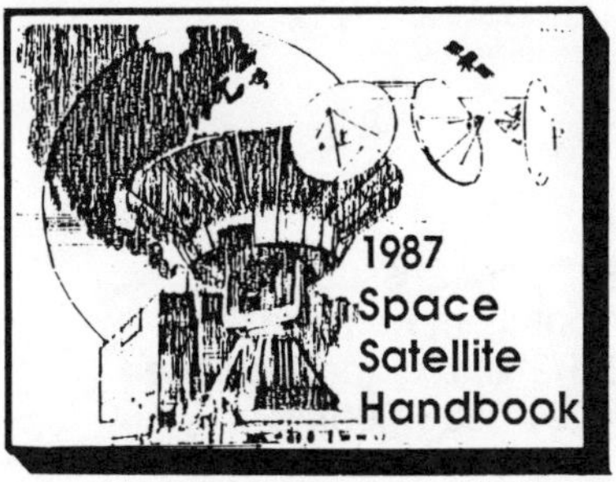

Space Satellite Handbook

LISTING ALL SATELLITES EVER IN ORBIT

edited by Anthony R. Curtis
editor of *SPACE TODAY* Magazine

This unique annual encyclopedia describes all 17,304 payloads, platforms, rockets and debris ever in orbit—from all countries.

A total of 6,237 man-made objects from spacefaring nations remain in orbit today from as far back as 1958. Some 11,067, no longer in orbit, are listed. This is the latest up-to-the-minute NASA, NORAD and Smithsonian Observatory data collected in an authoritative guidebook.

Each satellite, platform, rocket or piece of debris is listed with its official international number, popular name, launch date and country of origin. Orbital parameters and technical information included in this huge softcover reference work are period, inclination, apogee, perigee, and NASA radio frequency. A *Country Scoreboard* section compares satellites from 19 nations and five international space organizations.

You can look up complete specifications on any satellite from the many spacefaring countries. You will know, instantly, history and orbit of any satellite ever in space. This exciting book is the ultimate resource for space-minded students, teachers, hobbyists, space buffs, astronomers, engineers, scientists, anyone needing to know about everything in orbit.

This giant reference includes Australia, Brazil, Canada, Czechoslovakia, France, Great Britain, India, Indonesia, Italy, Japan, Mexico, The Netherlands, Peoples Republic of China, Saudi Arabia, Spain, Sweden, USA , USSR, West Germany.

Payloads, platforms, rockets, miscellaneous debris, all are in this space-buff's bible. Debris includes hatch covers, nosecone shrouds, even camera lenses remaining in orbit after payloads and rockets disintegrate or left behind by men in space. If it's big enough to be detected on Earth, it's included in this giant handbook.